SYSTEMATIC BOTANY MONOGRAPHS

VOLUME 65

Systematics of Austromyrtus, Lenwebbia, and the Australian species of Gossia (Myrtaceae)

Neil Snow
Gordon P. Guymer
G. Sawvel

THE AMERICAN SOCIETY OF PLANT TAXONOMISTS
19 May 2003

SYSTEMATIC BOTANY MONOGRAPHS
ISSN 0737-8211

ISBN 0-912861-65-7

Printed in the United States of America

SYSTEMATICS OF AUSTROMYRTUS, LENWEBBIA, AND THE AUSTRALIAN SPECIES OF GOSSIA (MYRTACEAE)

Neil Snow
Department of Biological Sciences
University of Northern Colorado
Greeley, Colorado 80639, U.S.A.

Gordon P. Guymer
Queensland Herbarium
Brisbane Botanic Gardens Mt. Coot-tha
Toowong 4066, Queensland, Australia

G. Sawvel
Department of Biological Sciences
University of Northern Colorado
Greeley, Colorado 80639, U.S.A.

ABSTRACT. *Austromyrtus* (Nied.) Burret is divided into three genera based primarily on morphological characters of the fruit and embryo. Two new genera, *Lenwebbia* and *Gossia*, are proposed and contrasted with *Austromyrtus* s.s. and *Myrtus*. Essential oil profiles of these and related genera are reviewed. A gelatinous endosperm occurs in some species of *Gossia*. "Specimen aggregation analysis" of herbarium specimens that bear fixed characters is presented as an approach to determine character fixation as a part of a diagnosability criterion under a general lineage species concept. The tripartite typological syndrome of embryos in fleshy-fruited genera of Myrtaceae should be abandoned, since embryos can be described by numerous character states. As emended, *Austromyrtus* includes *A. dulcis*, *A. tenuifolia*, and the newly described *A. glabra*. *Lenwebbia* comprises *L. lasioclada* and the newly described *L. prominens*. Most species formerly placed in *Austromyrtus* are now assigned to *Gossia*, which in Australia includes twenty species. Of these eight are newly described: *G. bamagensis*, *G. byrnesii*, *G. grayi*, *G. lewisensis*, *G. macilwraithensis*, *G. punctata*, *G. retusa*, and *G. sankowskiorum*. Twelve new combinations are proposed: *G. acmenoides*, *G. bidwillii*, *G. dallachiana*, *G. floribunda*, *G. fragrantissima*, *G. gonoclada*, *G. hillii*, *G. inophloia*, *G. lucida*, *G. myrsinocarpa*, *G. pubiflora*, and *G. shepherdii*. Lectotypes are designated for the basionyms *Myrtus acmenoides*, *M. fragrantissima*, *M. hillii*, *M. inophloia*, *M. racemulosa*, *M. shepherdii*, and *M. tenuifolia*. All species are mapped and the newly described ones are illustrated. Keys are provided for all genera (for *Gossia* separate keys for fertile and sterile material.)

INTRODUCTION

The Australian species treated here were mostly first described in *Myrtus* L. (Myrtaceae) and later transferred to *Austromyrtus* (Nied.) Burret (Bentham 1867; Niedenzu 1893; Bailey 1900; Burret 1941; Byrnes 1982; Stanley & Ross 1986; Henderson 1997; Hyland et al. 1999). As explained below in the section Generic Boundaries, circumscriptions of *Myrtus* and *Austromyrtus* have been unstable. We continue to recognize *Austromyrtus*, but in a narrowed sense, and propose the genera *Lenwebbia* and *Gossia* to accommodate some species formerly placed in *Austromyrtus* (Stanley & Ross 1986; Hyland et al. 1999) and others newly described herein.

Among the subtribes traditionally recognized under subfamily Myrtoideae (Candolle 1828; McVaugh 1956a; Landrum & Stevenson 1986; Landrum & Kawasaki 1997), *Austromyrtus*, *Lenwebbia*, and *Gossia* are clearly assignable to subtribe Myrtinae, based on the hard outer testa of the seed and on a number of characters of the embryo. Emerging DNA sequence data (Wilson et al. 2001; Salywon & Snow, unpubl.) increasingly suggest that the traditionally recognized subfamilies Leptospermoideae (dehiscent capsular fruits) and Myrtoideae (indehiscent berries), as well as some tribes and subtribes (e.g., Myrtoideae subtribes Eugeniinae, Myrciinae, and Myrtinae), are paraphyletic or polyphyletic (Briggs & Johnson 1979).

In its newer, more narrowly defined sense, *Austromyrtus* is a genus of three species in northern New South Wales and southeastern Queensland. Species of *Austromyrtus* are somewhat prostrate to sprawling or erect shrubs, which can be quite common locally. They are immediately distinguished in the family by the whitish to grayish mature fruits bearing numerous small dark spots, which often are somewhat pendulous on solitary peduncles. Species can be recognized vegetatively on close inspection by the bark of the branchlets (=growth of current year), which soon peels into four distinct strips (Fig. 3). In most cases, the leaves of *Austromyrtus* are generally smaller than those found in species of *Lenwebbia* and *Gossia*.

Lenwebbia consists of two species in northern New South Wales and southeastern Queensland, one of which also has a disjunct occurrence in northern Queensland. The species are rare to locally common. Mature specimens are nearly always trees, although younger individuals can be somewhat shrubby. The fruits are always dark purplish or black at maturity and are held erect. The generic characters that separate this genus from other fleshy-fruited genera in Australia are somewhat cryptic, and include a 3–4-loculed ovary, an apically to subapically attached lamelliform placenta, a C-shaped embryo, and a sunken midrib (at least proximally) on the adaxial leaf surface (Figs. 6, 7).

Gossia is the largest genus treated herein and has a bimodal distribution in eastern Australia and New Caledonia (Snow 2000, unpubl.). In Australia it consists of twenty species and ranges from northern New South Wales to the northern tip of the Cape York Peninsula in Queensland. *Gossia floribunda* also extends into New Guinea. Approximately fifteen additional species occur in New Caledonia, including some that are undescribed. Members of *Gossia* are shrubs to tall trees and very rare to locally abundant. The generic characters that uphold the recognition of *Gossia* are less easily recognizable than those of *Austromyrtus* and *Lenwebbia*. Mature fruits of *Gossia* are usually dark purplish to black and held on erect peduncles. Many species have greater amounts of branching in the inflorescence, as compared to *Austromyrtus* and *Lenwebbia*. The embryos of most species of *Gossia* are strongly circinate (coiled) and in eleven species are associated with small amounts of gelatinous endosperm (e.g., *G. bidwillii*).

None of the species treated herein are considered to provide useful timbers (Floyd 1989), but many are planted as attractive ornamentals. For example, *Austromyrtus tenuifolia* and *A. dulcis* are used horticulturally (Wrigley & Fagg 1996). Several species of *Gossia* are cultivated in home gardens in Queensland (Snow, pers. obs.; A. Ford, pers. comm.). Successful propagation of species of *Gossia* from cuttings can be difficult (A. Ford, pers. comm.). The "python bark" species of *Gossia* (*G. bidwillii*, *G. acmenoides*, *G. lucida*, and *G. grayi*) have a very smooth bark irregularly covered with relatively large, colored patches. The term "python" refers to the resemblance of the bark coloring to the skin coloring of the python snake, and although apparently not commonly grown, these species would make attractive additions to private and public gardens. Although the two species of *Lenwebbia* probably have the least horticultural appeal, they have been successfully grown in home gardens by botanists in Queensland (Snow & Guymer, pers. obs.). Additional horticultural value and suggestions for successful propagation of many species here discussed can be found in Jones (1986).

We redefine *Austromyrtus* and propose the new segregate genera *Lenwebbia* and *Gossia*, and also asssess the position of these three genera among several morphologically similar ones. Priorities for future studies include the revision of New Caledonian species of *Gossia* (currently undertaken by the first author) and the use of molecular markers to test hypotheses of phylogenetic relatedness within and among genera of Myrtinae (Snow 2000; Wilson et al. 2001; Salywon et al. 2002). Following the caveats of Doyle (1992) and others, however, a general congruence of morphological and molecular data sets should be achieved before significant generic re-alignments and the attendant nomenclatural changes are proposed.

TAXONOMIC HISTORY

Generic differences among fleshy-fruited members of Myrtoideae and its historically constituent subtribes (Eugeniinae, Myrciinae, and Myrtinae) have been based on cryptic characters of the inflorescence, flower, fruit, and embryo (Candolle 1828; Berg 1855–1856; Burret 1941; Kausel 1956; Wilson 1960; Hyland 1983; Landrum 1986, 1988; Landrum & Stevenson 1986; Gentry 1993; Landrum & Kawasaki 1997; Snow & Guymer 1999, 2001a). The continued uncertainty of generic boundaries, and their recognition based on cryptic characters, can largely account for the longstanding uncertainty regarding the supraspecific treatment of Australian members of *Austromyrtus* (Bentham 1869; McVaugh 1963, 1968, 1969; Briggs & Johnson 1979).

Australasian species originally described in *Myrtus* and treated herein have had an unstable taxonomic history above the species level. Landrum (1988) and others have emphasized the need for revisionary studies in Australasian Myrtinae before more concise hypotheses of generic boundaries could be tested, particularly those regarding the generic circumscription of *Myrtus*.

Most of the previously described species treated herein were first placed in *Myrtus*. Over one thousand names exist in *Myrtus*, and in past times the genus has been circumscribed far more broadly and included species from throughout the geographical range (mostly tropical) of the family (Candolle 1828; Berg 1855–56; Hooker 1864; Bentham 1867; Niedenzu 1893; Bailey 1900; Domin 1928; Guillaumin 1938; Burret 1941). Smith (1797) was the first to describe a species of the group here treated in *Myrtus*, namely *Myrtus tenuifolia*, which McVaugh (1956b) used to typify *Austromyrtus*; several more Australian species were described in *Myrtus* between 1788 and 1892 by authors such as Gaertner, Bentham, and Mueller. Niedenzu (1893) divided *Myrtus* into the subgenera *Calomyrtus* and *Pseudocaryophyllus*. *Calomyrtus* was divided further by Niedenzu into his newly created sections *Eumyrtus* and *Austromyrtus*. Between 1900 and 1939, Australian workers, such as Bailey and White, continued to describe new species in *Myrtus*.

Some fifty years after the work of Niedenzu, Burret (1941) elevated *Myrtus* sect. *Austromyrtus* Nied. to the generic level as part of a revision of fleshy fruited genera. He proposed many new combinations under *Austromyrtus*, including a number for species from Australia and New Caledonia. Since Burret did not provide a key to separate the genera, and his Latin descriptions do not include diagnoses contrasting related genera, it is not always obvious which characters he considered most important in separating the genera. For example, with regards to *Austromyrtus*, he indicated only that its species could be distinguished from those of *Myrtus* by their "*Eugenia* placentation," and from *Eugenia* by their "*Myrtus* seeds." These terms are an example of typological thinking in the systematics of Myrtaceae, which has often impeded progress in proposing a useful taxonomy for the family (see below). Burret's concept of *Austromyrtus* corresponded fairly closely to the three genera treated herein, although some generic transfers have been made (Scott 1986).

Most workers since the time of Burret (1941) have accepted *Austromyrtus* as a valid genus (Smith 1956; Floyd 1973; Briggs & Johnson 1979; Johnson & Briggs 1985; Wilson 1991; Stanley & Ross 1986; Guymer 1988; Hyland et al. 1999); however, others questioned whether *Austromyrtus* merited separation from *Myrtus*, and continued to describe or recognize species in the latter genus (Scott 1979, 1980, 1986; Byrnes 1982). Rationales for the separation of these genera have relied on reproductive characters, such as type and placement of placentation, number of locules, number and placement of ovules, and characters of the inflorescence.

Our taxonomic separation of *Austromyrtus*, *Lenwebbia*, and *Gossia* from *Myrtus* effectively ends a 250-year historical process in which the generic boundaries of *Myrtus* increasingly have been narrowed. We thus concur with those who recognize *Myrtus*, the type genus of Myrtaceae, as a single but highly polymorphic species native to the Mediterranean region.

MATERIALS AND METHODS

Species recognition. We recognize species that are invariant (fixed) for at least one morphological character or character state. Each of the 22 hypothesized "new" species of *Austromyrtus* in Henderson (1997) was examined for evidence of character state fixation in the context of previously described species in *Austromyrtus*, *Lenwebbia*, and *Gossia*.

As with most taxonomic revisions of vascular plants, herbarium specimens represented the majority of study material. As such, we did not delimit species using "population aggregation analysis" (PAA) (Davis & Manos 1991; Davis & Nixon 1992); rather, we aggregated herbarium specimens based on character fixation into hypothesized species using "specimen aggregation analysis" (SAA), a process outlined earlier but not formally named (Snow 1997). As a methodology for determining species boundaries based on morphology, SAA is operationally similar to PAA in its reliance on the diagnosability criterion of fixed characters (Davis & Manos 1991; Davis & Nixon 1992); however, whereas PAA relies on character fixation at the population level, SAA uses the invariability of a character or some combination thereof on herbarium specimens as the basis for species recognition (Snow 1997). (This is not to suggest, however, that population-based studies should not be undertaken to the maximum practical extent during the revisionary process.) Thus, if character state x consistently occurs on all specimens, and the sample of specimens includes a representative from at least several different populations, then it is reasonable to assume that x is also fixed at the population level and represents a reliable diagnostic character for that species (see also Blake, 1949). If additional study revealed x to be polymorphic, then its reliability as a diagnostic character for that species should be reassessed (Davis & Nixon 1992; Frost & Kluge 1994; Snow 1997). Wiens and Servedio (2000) recently discussed the difficulty of demonstrating character fixation in species (apart from those that are known from one or a few small populations), and presented statistical approaches to help assess whether fixation of characters is a reasonable assumption based on sample size and other features.

Over 1,400 herbarium specimens were examined for this study (see Acknowledgments). Apart from *Gossia bamagensis*, *G. grayi*, and *G. lucida*, living material has been observed for all species by either the first or second author. Supplemental observations of living material have been made by Andrew Ford of Atherton for several species in *Gossia* and communicated to the first author. Personal observations by other botanists of living specimens of *G. bamagensis*, *G. grayi*, and *G. lucida* are noted as appropriate following the species descriptions.

Maps of the taxa are meant to portray the overall geographical range and approximate the density of collections within the range. In some cases it was necessary to remove collecting points to avoid cluttering the maps.

Character coding. Some 140 morphological characters were analyzed for variation, building upon and modifying an earlier set of characters for fleshy-fruited Myrtaceae (Snow & Guymer 1999; see Morphology below). Terminology regarding shapes generally follows the recommendations of the Systematics Association Committee (1962). Aspects of leaf architecture generally follow Hickey (1973). Terminology of hair characters generally follow Hewson (1988), although "sericeous" replaces the term "silky." Several characters of the inflorescence absent from the generic descriptions were assessed, given their discussion by Briggs and Johnson (1979) relative to the recognition of generic and suprageneric taxa in Myrtaceae. Characters and character states were coded in DELTA (Dallwitz et al. 1993) and their ability to separate taxa was further explored using Lucid software (CPITT 1999). Some characters invariant within a genus are indicated in the generic descriptions in light of their likely value in broader studies of Myrtaceae. For example, the entire embryo is indicated as being held on a single horizontal plane, in contrast to some species of *Lithomyrtus* F. Muell., in which the radicle end of the embryo is spirally contorted above the plane of the embryo (Snow & Guymer 1999). Upon publication of this work, a Lucid™ based interactive key Australian species of *Gossia* will be posted on the herbarium website of the University of Northern Colorado.

Essential oils. The majority of species in *Austromyrtus*, *Lenwebbia*, and *Gossia* have been examined for secondary chemical compounds (Brophy et al. 1995a, b, c, 1996a, b, 1997a, b; Brophy, unpubl.). Data from these sources are reviewed below.

Specimen citations. In this first comprehensive treatment of Australian material of *Austromyrtus*, *Lenwebbia*, and *Gossia*, which includes several species that are of immediate conservation concern, sterile specimens are sometimes cited (Snow & Keating 1999). Descriptions for species in the rainforest region of northern Queensland have been compared to information in Hyland et al. (1999).

MORPHOLOGY

The delineation of character states is perhaps the most difficult aspect of taxonomic description with respect to considerations of subjectivity and repeatability (Gift & Stevens 1997). The descriptions of characters provided below apply to the three genera, except as noted. Many species treated herein are newly described and others described in detail for the first time. As such, the choice of terms employed for characters and character states primarily reflects their value as descriptors. Their value for purposes of phylogenetic reconstruction is a secondary consideration at the present time, since species (described and undescribed) of *Gossia* from New Caledonia must be examined critically before character states in that genus can be described with the primary goal of using them as phylogenetic markers. Many descriptors, such as the texture of surfaces (e.g., bark), colors, and relative abundance of oil glands and hairs, should not be used for phylogenetic inference. With certain exceptions where noted (e.g., shape of cross section of stem of seedlings), character states apply to plants at reproductive maturity.

Habit. All species of Myrtaceae are woody. The three species of *Austromyrtus* are sprawling to erect shrubs, whereas species of *Lenwebbia* and *Gossia* collectively range from short, multi-stemmed shrubs to large trees. Main trunks are generally round, although buttressing or fluting occurs in a few species of *Gossia*, particularly those of the wettest rainforests in northern Queensland.

Bark. The texture of the bark can vary depending on the age of the individual and the time of year; however, the texture of each species generally is consistent enough to be described as smooth or variously flaky, fissured, tessellated, etc. The smooth bark on some species in *Gossia* is unmistakable and may be a synapomorphy uniting some taxa (e.g., *G. bidwillii* and *G. acmenoides*). The term "python bark" often is applied informally to species with a smooth bark that bears irregular patches of various sizes and colors. Our descriptions of bark texture have been reviewed in light of observations of Hyland et al. (1999) for species from the rainforests of northeastern Queensland.

In all three species of *Austromyrtus* the bark of branchlets soon peels into four distinct strips (Fig. 3). This generic character may be neotenic (Alberch et al. 1979); bark has been observed to peel away similarly in seedlings of *Gossia bidwillii* (*Gray 2319*, QRS).

Bark color also varies depending on the age of the individual. Blake (1949) cautioned about the variability of bark color throughout the year; still, bark color can be a useful diagnostic character in the field. Descriptions of bark color have been reviewed in light of observations of Hyland et al. (1999) for species from the rainforests of northeastern Queensland.

Branchlet shape. Branchlets (growth of current year) are generally round. Internodes can be flattened near the apex but rounded near the base. Some species are clearly quadrangular or round with four evident extensions (e.g., *Gossia retusa*, Fig. 24). Other species have two narrow latitudinal wings arising on each side near the somewhat flattened apex of an internode (e.g., *G. gonoclada*), giving a superficial resemblance to a quadrangular stem. Seedlings of *G. bidwillii*, *G. hillii*, and many other species are four-angled or winged, in contrast to mature specimens of each species (Hyland et al. 1999). This pattern recurs frequently in the seedlings of many species of Myrtaceae (Snow, pers. obs.). Thus, characters of the stem may change during ontogeny, and the diagnostic features listed in the descriptions apply only to older plants, except as noted.

Pubescence. Terminology of pubescence largely follows Hewson (1988). Young leaves are generally sericeous but soon become glabrous. Pubescence characters apply to fully developed leaves, unless noted otherwise.

Stipules. Weberling (1956, 1984, 2000) discussed the presence of stipules in Myrtales and Myrtaceae. We concur that the hairlike projections at the base of of the petiole (and sometimes near the outer edges) are stipules, although we have not used the term "false stipules" (e.g., Weberling 2000) for the expanded edges of four-angled internodal apices when they are present. Assigning character states to stipule types in Myrtaceae and the genera treated here is somewhat arbitrary.

The most common character state consists of a few setose, hairlike structures, which are always thicker and longer than any pubescence on the adjacent stem. In most cases the setose hairs are ferrugineous, but they also can be whitish (e.g., *G. fragrantissima*). We abbreviate this description as a "type A" stipule. A modification of the type A stipule is prominently swollen at the base (e.g., *G. dallachiana*, *G. floribunda*), which is designated "type B." In *G. macilwraithensis*, the hairlike stipules extend laterally between the petioles on the branchlets, resembling interpetiolary stipular hairs. In "type C," stipules are scalelike and resemble small, undifferentiated axillary buds (e.g., *G. shepherdii*, *G. myrsinocarpa*). In *G. lewisensis*, short, faint (often clear) ochreate stipules, designated "type D," exist on the younger nodes of branchlets but soon deteriorate. Stipules are generally somewhat obscure and require magnification to be seen.

Petioles. The adaxial surface of the petiole can be flat to slightly indented or deeply channeled. This feature is measured at the midpoint of the petiole, unless indicated otherwise. It can be an excellent diagnostic feature in the field; for example, the presence of a prominent channel in *G. bidwillii* and its absence in the closely related *G. acmenoides*.

Petioles may have distinct, often large oil glands when young, but these tend to become obscured or disappear as the petiole becomes increasingly sclerified in age.

Leave blades. Leaf shape can vary considerably within a species. Some specimens may have a few leaves with a shape that is considerably atypical. In these cases, the blades are often highly asymmetrical and somewhat falcate (e.g., *G. lucida*, *Fell DGF4317*, BRI).

Leaf "apex" and leaf "tip" are approximate terms, but it is useful to differentiate between them (Fig. 1). The apex designates the distal 10–25% of the blade (Hickey 1973), whereas the tip indicates the distal-most portion of the apex. Distinguishing between these regions is necessary to prevent confusion, because an acute to acuminate apex can have a rounded or pointed tip.

Descriptions of leaf margins are based on dried material, although the margin sometimes becomes more recurved at the edges after drying than is typical in living plants. *Gossia retusa* consistently has recurved margins, which is also common in species of *Austromyrtus*, particularly near the base of the blade. *Gossia fragrantissima* is the only species treated here that regularly has somewhat undulate margins.

Venation in the genera is brochidodromous (Hickey 1973). The veins that connect in a series of the outermost loops of the secondary veins (Hickey 1973) are generally referred to as "intramarginal" veins in Myrtaceae (Fig. 1). An "outer intramarginal" vein, which typically is much more faint, can occur distal to the "inner intramarginal" vein, which is always more prominent. Some species can have three intramarginal veins (e.g., *Gossia dallachiana*), in which case the innermost prominent intramarginal vein is measured, even if its origin is distal to the junction of the base of the leaf and the petiole.

Klucking (1988) also examined leaf venation patterns of four species of "*Austromyrtus*": *Gossia bidwillii*, *G. dallachiana* (as *A. delachayii*), *G. myrsinocarpa* (as *A. minutiflora*), and *Lenwebbia prominens* (as *A. lasioclada*). He assigned these taxa to four groups based primarily on characters of secondary venation and secondarily of intercostal venation (Klucking1988: 246). Although these groups may help highlight variation in venation patterns, each is a composite of several features and is thus typological (see below). In

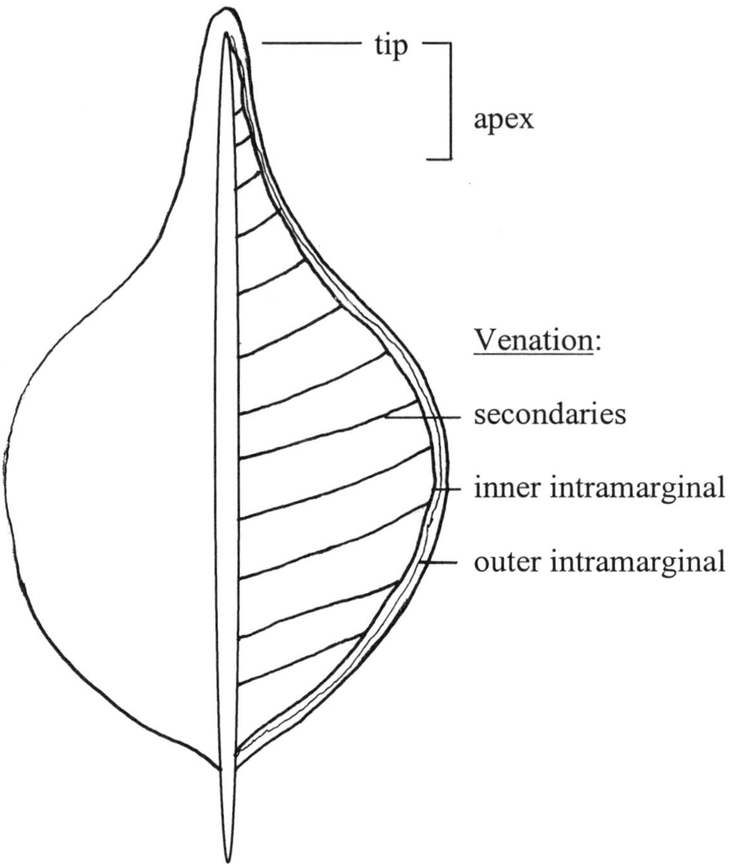

FIG. 1. Diagram of a leaf showing regions defined by "apex" and "tip." The inner intramarginal vein typically is relatively prominent and connects directly to secondary veins arising from the midvein; the outer intramarginal vein typically is faint and located between the inner intramarginal vein and leaf margin.

addition, duplicates of specimens cited by Klucking for *G. myrsinocarpa* (*Gentry 34417*, *34453*, *34354*) seen at MO are sterile, and their identity remains uncertain.

Observations of leaf reflectance (sheen) are based on dried specimens, which may differ somewhat from fresh material. Sheen ranges from very shiny on the adaxial leaf surface of *G. hillii* to matte in most species.

Leaf oil glands. Terms indicating glandular presence are based on the spacing of glands, as based on glandular diameters: abundant (0–2× diameter); common (2–5× diameter); sparse (>5× diameter). Reports of glandular prominence (not visible with the naked eye, prominent, etc.) are based on observations made with a dissecting scope with 50× magnification and an attached front-lit light of medium to high intensity. Oil glands are generally seen as more prominent when backlit.

Inflorescence. The complex inflorescence architecture of Myrtaceae was reviewed and described comprehensively by Briggs and Johnson (1979). Only dedicated specialists in the family can fully understand this terminology, and it generally has not been followed in many of its details (e.g., Landrum 1988).

Difficulties can arise when trying to describe the inflorescences of *Gossia*. For example, following Briggs and Johnson (1979), the short flowering axis in the axils of leaves in *G. floribunda* can be interpreted as either a short shoot from which numerous inflorescences of monads arise (our interpretation) or as a botryoid; however, the terms panicle, raceme, and solitary generally can be applied to the inflorescence if allowance is made for the fact that bracts of various sizes (usually less than 5 mm in length) can occur within the inflorescence. The following terms used by Briggs and Johnson are indicated with our equivalents in parentheses: thyrse (panicle), dichasium (panicle), botryoid (raceme), monad (solitary flower). Triad (=simple dichasium) is used in the sense of Briggs and Johnson (1979: 179, their figure 3b–d), but we use "dimer" to refer to a pedicel bearing two flowers, as if the central flower were missing in figure 3d of Briggs and Johnson (1979: 179).

The appearance of the inflorescence can change between flowering and fruiting periods. Sometimes lateral flowers on short floral axes break off prior to fruiting, which without close inspection erroneously suggests the fruits occur on unbranched pedicels.

Inflorescence variation can be pronounced, even on an individual plant. For example, a thyrse (panicle), dichasium (panicle), and a botryoid (raceme) all arise from a single axil in the spirit material of the type specimen of *G. punctata* housed at BRI, and monads occur elsewhere on the same specimen.

According to an analysis of putative homologies, the ancestral inflorescence in Myrtaceae may have been a single flower (monad). Stipular structures (hairs and their modifications) generally occur in the axils of the bracteoles, which, given their occasional midvein and leaflike appearance, are probably ontogenetically individualized (sensu Wagner 1989) derivatives of the leaf.

As used here, peduncle refers to the axis below a flower, whether solitary or as part of more complex inflorescence, on which anthopodia and/or metaxphylls (sensu Briggs & Johnson, 1979: 245–246) may exist. For caveats regarding the use of the term peduncle more generally in Myrtaceae, see also Briggs and Johnson (1979).

All species treated here generally have two bracteoles that subtend the flower, but these often fall early in anthesis. When bracteoles are indicated as absent (e.g., *G. byrnesii*, *G. bamagensis*) a very few may be present, but since they are lacking in the overwhelming majority of flowers, some genetic repression of bracteole expression apparently exists. "Foliose" refers to bracts that resemble miniature leaves, given laminar-like lateral extensions, whereas "scalelike" reflects the lack of such extensions (Fig. 2).

"Anthopodium" refers to a naked axis between the base of the flower and the bracteoles (Briggs & Johnson 1979, Fig. 3; 179, pp. 241); this structure is not common in taxa treated here (Fig. 2).

The term "metaxyphylls" refers to a pair of sterile, relatively small bracts beneath the bracteole; such bracts are not common in taxa treated here (Fig. 2). Variation in this feature can occur within species and even on individuals. For example, separate inflorescences on a single specimen of *G. bidwillii* (e.g., *Bravery s.n.* [QRS 3090]) have flowers with and without metaxyphylls.

Hypanthium. Most species have a cupulate hypanthium, i.e., the hypanthium (with included inferior ovary) and the adnate bases of the calyx lobes resemble a bowl (Fig. 2). Yet, in some species, such as *G. retusa* and *G. sankowskiorum*, the hypanthium is

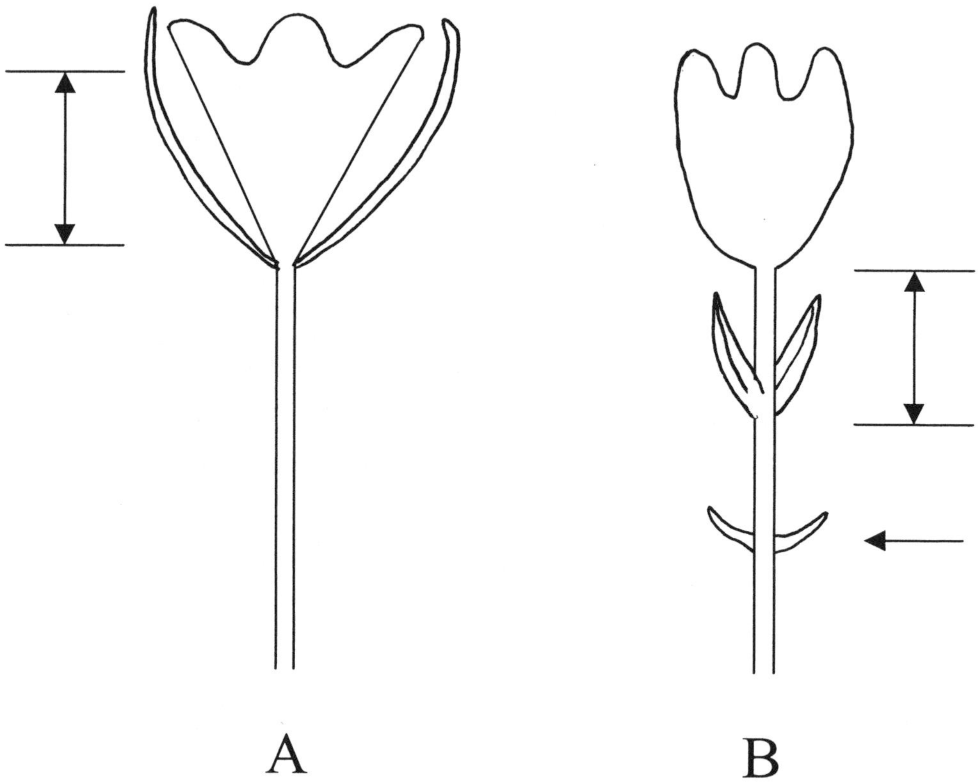

FIG. 2. Floral morphology. A. Obconic hypanthium; tips of scalelike bracteoles exceed base of sepal lobes. The two-headed arrow indicates the length of the hypanthium. B. Cupulate hypanthium; the anthopodium is located in the region indicated by the two-headed arrow; the horizontal arrow points to the place of insertion of metaxyphylls. (Terminology follows Briggs and Johnson, 1979.)

distinctly obconic with a clearly tapered base. Whether this distinction holds value beyond the taxa treated here is uncertain.

The length is measured from the base of the hypanthium to the base of the calyx lobes. The base of the hypanthium begins where the hypanthium equals the diameter of the axis of the peduncle.

Calyx and corolla parts. *Austromyrtus* is 5-merous, *Lenwebbia* 4-merous, and *Gossia* (3–) 4–5-merous. These conditions are usually consistent within a species, but are known to vary in *Gossia* and other genera of Myrtaceae, such as *Pimenta* Lindl. and *Myrteola* O. Berg (Landrum 1988; Landrum & Kawasaki 1997).

Stamens. The number or rows (=series) of stamens can vary on a given flower. For example, in *G. retusa* (*Morton 1007*, QRS), two or three series occur directly opposite the calyx lobes, whereas between the lobes the number rises to four or five. The number of stamens counted in bud can be lower than when counted in opened flowers, which suggests stamen initiation and development may occur until late in floral ontogeny.

The mode of anther connection to the filament is generally consistent within a species, but can be both basal and dorsifixed in *G. bidwillii* and *L. lasioclada*. The basal attachment of the connective to filament still allows versatility, because the base of the connective is considerably shorter than the anther sacs. Oil glands visible on an anther in bud may be nearly invisible (e.g., *G. dallachiana*) or prominent during anthesis (e.g., *G. sankowskiorum*).

Placenta. Placenta structure varies considerably in Myrtaceae. Some have questioned whether placental characters can reliably differentiate *Austromyrtus* sensu lato from *Myrtus* (Byrnes 1982; Scott 1979). This concern may reflect the fact that the appearance of the locules and placenta can change during ontogeny. In *Myrtus communis* L. the locular wall generally extends only partially towards the top of the ovary in the mature fruit (Snow, pers. obs.; Landrum, pers. comm.). In contrast, the locular wall in the mature fruit of species in *Austromyrtus*, *Lenwebbia*, and *Gossia* generally is entire (Snow, pers. obs.). In *Austromyrtus* and *Gossia*, the placenta typically is more or less capitate and located near the middle of the locule. In *Lenwebbia* the placenta is lamelliform with an apical or subapical attachment (Fig. 6c).

Ovule number. The number and placement of ovules varies and is often taxonomically useful at the species level. In *Austromyrtus*, *Lenwebbia*, and *Gossia* there is typically little obvious patterning to the placement of ovules on the placenta, unlike some genera in Myrtaceae.

Fruit color. The color varies according to age and whether fresh or preserved. Younger and fresh fruits in *Gossia* and *Lenwebbia* are generally green and become increasingly darkened as they mature.

Seeds. The seed coat (outer testa) is generally bony and quite hard; however, it may be very thin and somewhat transparent (e.g., some specimens of *G. bidwillii*, such as *Guymer 1125*, BRI) or somewhat leathery.

Hard seed coats often have a more or less distinct, circular operculum, which opens upon germination (Landrum & Kawasaki 1997: 514). In some cases the feature is obvious; in others, it is evidently absent.

Embryo. Embryo shape appears to be a reasonably consistent character at the specific and generic levels (see also Landrum & Stevenson 1986; Landrum 1986). The circinate embryo here hypothesized as a character state synapomorphy for species of *Gossia* can occur elsewhere in Myrtaceae, as for example *Campomanesia* Ruiz & Pav. (Landrum & Kawasaki 1997; Kawasaki 2000) and *Blepharocalyx salicifolius* (H. B. K.) O. Berg (Landrum & Kawasaki 1997). Embryo shape occasionally changes during development. For example, the mature embryo is circinate in *Gossia bidwillii* (e.g., *Gray 2319*, BRI), but younger embryos can be S-shaped (Snow, pers. obs.).

Embryo oil glands were studied earlier in Myrtaceae by Petit (1908). Although we indicate the presence or lack of oil glands in the species descriptions, this feature seems to be of limited diagnostic value; its expression seems to vary during embryo ontogeny. Also, light conditions and the relative dryness of the embryo at the time of viewing affect visibility. Also, glands on the embryo can be the same color as surrounding tissues and therefore nearly invisible.

The radicle is here reported as relatively long or relatively short, in comparison to the hypocotyl. Landrum (pers. comm.) considers the radicle to be a very inconspicuous part of the very tip of the hypocotyl in embryos of subtribe Myrtinae. If a radicle (tip of hypocotyl) is clearly much thicker than the accompanying cotyledons, then it is reported as thick. If it is about the same thickness as the cotyledons (together), it is considered thin. The radicle apex is reported as swollen near tip or not (or barely) swollen.

The cotyledons typically are held straight in the seed, but can be circinate. They are either about the same length as the hypocotyl or considerably shorter. The tips of cotyledons typically are pointing apically forward (away) from the hypocotyl or less frequently are folded backwards towards the hypocotyl. Their thickness is measured relative to the hypocotyl; typically, the hypocotyl is considerably thicker than the cotyledons.

Endosperm. Eleven species of *Gossia* were found to include at least some specimens that bear small amounts of a gelatinous endosperm (clear and colorless or somewhat milky) surrounding a portion of the embryo. For example, a collection of *G. bidwillii* from Bowen in Queensland (*Guymer s.n.*, BRI, in spirits) has embryos with a prominent band of a clear, colorless, gelatinous endosperm occuring between portions of the circinate embryo (Snow, pers. obs.). Other workers have reported a gelatinous endosperm in embryos of Myrtaceae (van Wyk & Botha 1984). A pencil drawing of embryos on the sheet bearing the holotype of the New Caledonian species *Myrtus randiana* Merr. & L. M. Perry (a species apparently belonging to *Gossia*) is accompanied by the observation "± gelatinous." As such, we are confident of our interpretation that a gelatinous endosperm occasionally is present in *Gossia*.

REJECTION OF TYPOLOGICAL DIVISION OF EMBRYO CHARACTERS

Typological thinking, in which an ahistorical class of objects is defined by some essence or archetype, has been criticized for its applications in biology on ontological and epistemological grounds (Hull 1965; Mayr 1982). Typological thinking has impeded progress in systematics from the perspective of taxon recognition (e.g., Gould 1989; Frost & Kluge 1994) and character recognition (e.g., Bradford 1998; Snow 1998; Floyd & Friedman 2000). Regarding character state recognition, what is today considered typological may have been considered, at the time of its description, the equivalent of what we recognize today as characters and character states. Thus, when Candolle (1828) recognized myrcioid, eugenioid, and pimentoid (=myrtoid) embryo types in subfamily Myrtoideae (see Landrum 1981; Landrum & Stevenson 1986), he may have recognized these as the equivalent of what we now call a single character (embryo) possessing three states.

Despite earlier suggestions to abandon typological thinking in Myrtaceae (McVaugh 1968; Legrand 1975; Briggs & Johnson 1979), many workers continue to use a tripartite typology to describe embryos of fleshy-fruited genera (e.g., Byrnes 1982; Landrum 1981; Landrum & Stevenson 1986; Landrum & Kawasaki 1997). In short, embryos are said to be eugenioid, myrcioid, or myrtoid. Although embryos of species in Myrtoideae can be classified easily as myrcioid, eugenioid, or myrtoid types, there are important variations within each of these archetypes. For example, Wilson et al. (2001) recently discussed how the cotlyedons are bent to one side of the curved hypocotyl in *Backhousia* Hook. & Harv. and *Choricarpia* Domin, two genera that form a monophyletic branch in their analysis of Myrtaceae. As such, the tripartite embryo conceptualization is abandoned here and replaced by hypothesized characters and character states (e.g., Grifo 1992; Snow & Guymer 1999; see above), which are more useful for testing hypotheses of homology and phylogenetic relatedness (Patterson 1982). The characters presented reflect the variation that occurs in fruit and embryo characters in *Austromyrtus*, *Gossia*, *Lenwebbia*, and represents an initial attempt to partition variation of embryos into characters and character states. The wider applicability of these characters in Myrtaceae remains to be determined. In contrast to Landrum (1981), we have found occasional intrageneric variation in embryos, particularly in *Gossia*, which further supports our suggestion that progress in understanding and testing generic relationships would be better served if embryonic archetypes in Myrtaceae were abandoned.

A final comment regarding typological thinking is pertinent relative to its value to the non-specialist user of monographs. The complex terminology of myrtaceous inflorescences advocated by Briggs and Johnson (1979), despite its usefulness in ontogenetic and phylogenetic applications, is only understandable to committed specialists in the family. We thus bypass much of that terminology for the benefit of those who would use our keys, and substitute typological terms, such as raceme and panicle (see also Landrum 1981; Landrum & Kawasaki 1997).

ESSENTIAL OIL PROFILES

The majority of species in *Austromyrtus*, *Lenwebbia*, and *Gossia* have been examined for secondary chemical compounds (Brophy et al. 1995a, b, c; 1996a, b; 1997a, b; Brophy, unpubl.). Except for restriction site fragment and DNA sequence analysis (e.g., Chappill & Ladiges 1996; Wilson et al. 2001; Salywon et al. 2002) the utility of most chemical compounds to help resolve generic relationships in Myrtaceae, and indeed angiosperms as a whole, has had limited success (Humphries & Richardson 1980). Still, secondary plant compounds, such as essential oils, can be useful in documenting similarities among taxa and, perhaps more compellingly, reveal variation within taxa. The following generalized discussion focuses on essential oils present in relatively high proportions in *Austromyrtus*, *Lenwebbia*, and *Gossia*.

The primary essential oils (by percentage; n = sample size) for species of *Austromyrtus* (Brophy et al. 1995a) were: *A. dulcis* (α-pinene 45.21%, 1,8-cineole 23.69%; $n = 3$), *A. glabra* (isobaeckeol 85.52%; $n = 3$), and *A. tenuifolia* (isobaeckeol 97.68%; $n = 3$). Interestingly, the predominantly coastal species *A. glabra* has a profile more similar to that of *A. tenuifolia*, which grows at higher elevations. (See Brophy and Doran, 1996, for illustrations of many chemical structures.)

Essential oils in species of *Lenwebbia* are predominantly sesquiterpenes (Brophy et al. 1995b). Of the collections analyzed in that study, only *Byrnes 4069* and *Forster PIF14614* are placed in *L. prominens*, whereas the remainder represent *L. lasioclada*. *Lenwebbia* has low levels of essential oils compared to *Austromyrtus* and *Gossia*. Considerable variation was detected in *L. lasioclada*. Samples of specimens from the Main Range region of Queensland (Moreton District) comprised alcohols and, to a lesser extent, unidentified sesquiterpene alcohols. Whereas the chemical profile of *Lenwebbia* is considerably at odds with that of *Austromyrtus* s. s., it is similar to that of the sesquiterpene-dominated group in *Gossia* (see below).

For *Gossia*, we have recognized as conspecific some specimens treated as distinct species by Brophy et al. (1995c, 1996a), viz.: *Gossia bidwillii* (=*Austromyrtus racemulosa*, "Danbulla," "Forty Mile Scrub," "Spencer Creek"); and *G. floribunda* (="Mt. White"). The dominant essential oil profiles allow partitioning of the species of *Gossia* into three groups (Brophy et al. 1995c, 1996). The first of these contains solely of *G. gonoclada*, which is dominated by 2-hydroxy-4,6-dimethoxy-3,5-dimethylacetophenone, a compound not found elsewhere in the genera revised herein. Sesquiterpenes dominate the essential oil profiles of a second group, including: *G. floribunda* (two of three chemotypes), *G. hillii* (one of two chemotypes), *G. myrsinocarpa* (= *A. minutiflora*), *G. pubiflora*, *G. bamagensis*, *G. punctata* (as "Brookfield" in Brophy et al.), *G. macilwraithensis*, and *G. bidwillii* (as "Pinnacle Track" in Brophy et al.). The third group is dominated by monoterpenes and includes: *G. acmenoides*, *G. bidwillii* (s. l., see above), *G. dallachiana*, *G. fragrantissima*, *G. hillii* (one of two chemotypes), *G. sankowskiorum* (="Byerstown Range"), *G. lewisensis* (="Mt. Lewis"), and *G. floribunda* (one of three chemotypes, as "Mt. White"). The Australian species of *Gossia* that remain untested for essential oils are *G. byrnesii*, *G. grayi*, *G. inophloia*, *G. lucida*, and *G. retusa*.

Given that *Myrtus communis* has been cultivated since ancient times for many purposes, including its aromatic properties, the taxonomic significance of its essential oils must be treated with caution. Brophy (pers. comm., unpubl.) studied a bulk sample of one specimen of *M. communis* and found that α-pinene (22%) and limonene (24%) constituted nearly 50% of its essential oils by weight (voucher: *Snow 7472*, BRI; Brisbane Botanical Gardens Mt. Coot-tha, Accession 8620415; seed source: Bordeaux, France). Brophy (pers. comm.) indicated that these results differ considerably from others published, which reiterates the need for a cautious interpretation of this widely cultivated species.

Among Australasian genera considered systematically close to *Myrtus*, α-pinene is also known in relatively high proportions (5% or more) from *Austromyrtus dulcis* (Brophy et al. 1995a), several species of *Gossia* (Brophy et al. 1995c, 1996a), *Uromyrtus metrosideros* (F. M. Bailey) A. J. Scott and *U. lamingtonensis* N. Snow & Guymer (2001) (as "McPherson Range" in Brophy et al. 1996b), four species of *Rhodomyrtus* (DC.) Rchb. (Brophy et al. 1997a), and at least nine species of *Rhodamnia* Jack (Brophy et al. 1997b). Limonene is also known from populations of *Lenwebbia lasioclada* from the Main Range region of Queensland (Moreton District) near Mt. Cordeaux and Wilson's Peak (Brophy et al. 1995b); in trace amounts from several species of *Gossia*; in proportions of up to 11.6 % in populations of *G. bidwillii* from the Forty Mile Scrub area of Queensland (Cook District) (Brophy et al. 1995c, 1996a); in trace amounts in species of *Uromyrtus* Burret (Brophy et al. 1996b), *Archirhodomyrtus beckleri* (F. Muell.) A. J. Scott (Brophy et al. 1996c) and species of *Rhodomyrtus* (Brophy et al. 1997a); and in trace to significant amounts in species of *Rhodamnia* (Brophy et al. 1997b).

Overall, data from essential oils do not help resolve generic boundaries among these and related genera. Many species in *Gossia* and *Lenwebbia* have sesquiterpenes as their most abundant essential oils, but this pattern is repeated in related the Australasian genera *Uromyrtus* (Brophy et al. 1996b), *Archirhodomyrtus* (Nied.) Burret (Brophy et al. 1996c), *Rhodomyrtus* (Brophy et al. 1997a), and *Rhodamnia* (Brophy et al. 1997b). The data do not contradict, however, the removal of *Gossia* and *Lenwebbia* from *Austromyrtus*, which we are proposing primarily on features of the fruit and embryo.

GENERIC BOUNDARIES

For many years, a number of fleshy-fruited species of Myrtaceae were placed in a broadly construed *Myrtus* (Candolle 1828; Berg 1855–56; Hooker 1864; Bentham 1867;

Niedenzu 1893; Bailey 1900; Domin 1928; Guillaumin 1938; Burret 1941). *Myrtus* has been a focal point regarding generic circumscriptions for American, Old World, and Australasian genera of fleshy-fruited Myrtaceae (McVaugh 1968, 1969; Byrnes 1982; Landrum 1989), and Landrum (1988: 120) stressed that "one of the greatest systematic problems awaiting resolution in the Myrtaceae is the definition of the limits of *Myrtus*." Uncertainty regarding generic boundaries between *Myrtus* and *Austromyrtus* has persisted until relatively recently (Scott 1979; Byrnes 1982). Our subdivision of *Austromyrtus* sensu lato here and description of the generic segregates *Lenwebbia* and *Gossia* corroborates the general consensus that *Myrtus* consists of a single species in southern Europe and the Mediterranean area, with possibly a second species occurring in the mountains of the Sahara (Briggs & Johnson 1979; Landrum 1989).

Austromyrtus is easily separated from *Lenwebbia*, *Gossia*, and *Myrtus* by its whitish gray fruits with dark spots and the condition of having the bark of the branchlets peeling in four distinct strips. It comprises species of a relatively low growth stature that are always shrubs.

Lenwebbia morphologically resembles *Austromyrtus*, *Myrtus*, and *Gossia*, but consistent differences in locule number, ovule placement, embryo shape, leaf venation (Klucking 1988), and other features are of sufficient magnitude to warrant generic status (McVaugh 1956a, 1968; Gentry 1993; Landrum & Kawasaki 1997). The lamelliform nature of the placenta is most evident in mature fruits, since the placenta shape in young buds can be somewhat capitate in early ontogeny. The somewhat angular seeds with the darker, softer centers are likewise atypical, although this seed morphology is not unique in Myrtaceae. *Lenwebbia* is similar to the New Zealand endemic *Lophomyrtus* Burret in the 4-merous flowers, a narrow and terete stigma, and the placement of numerous ovules on apical or subapical lamelliform placentas. *Lenwebbia* differs in having 3–4 locules with the ovules irregularly attached on the placenta, whereas *Lophomyrtus* has 2–3 locules with ovules uniseriate along the placental margins. DNA sequence data from the nuclear ribosomal DNA regions ITS-1 and ITS-2 of some ninety species in thirty genera of Myrtaceae support the hypothesis that *Lenwebbia* and *Lophomyrtus* are sister genera (Salywon et al. 2002, unpubl.).

Gossia differs from *Myrtus* by its circinate embryos (in most species). In addition, *Myrtus* is separated from *Gossia* by the much more tapered base of the fruit, the apically incomplete locular walls, the much greater relative length ratio of filaments to anthers in *Myrtus*, and the larger number (in general) of stamens. Many species of *Gossia* have relatively small amounts of a gelatinous endosperm adjacent to the embryo

Species in *Gossia* are grouped artificially here primarily on the basis of the number of perianth parts. Eight species have predominantly 5-merous flowers: *G. acmenoides*, *G. bidwillii*, *G. hillii*, *G. inophloia*, *G. pubiflora*, *G. lewisensis*, *G. punctata*, and *G. gonoclada* (which rarely also has 4-merous flowers). Secondary grouping in the 5-merous species was based on characters of the bark (always very smooth versus more or less rough), pubescence of the branches, and projection of the upper midrib of the leaf. The remaining species in *Gossia* are predominantly 4-merous (rare 5-merous exceptions occur in *G. floribunda*, *G. shepherdii*, and *G. lewisensis*). Secondary grouping of the 4-merous species was more difficult than of the 5-merous species, since many character states have rare to occasional exceptions from the most common state in a given species. Ignoring such polymorphisms would simplify the construction of keys and the formation of informal groups, but is unacceptable to ignore variation in comprehensive revisionary and phylogenetic work (Nixon & Davis 1991). Since these species groups do not reflect cladistic analysis they are likely artificial. Phylogenetic relationships can only be assessed meaningfully after the New Caledonian species that should be included in *Gossia* have been revised.

Landrum (1988) discussed similarities of New and Old World genera to *Myrtus* in a systematic treatment of *Myrteola*. Recent work on Australian species of *Rhodomyrtus*, *Archirhodomyrtus*, *Pilidiostigma* Burret, *Uromyrtus*, and *Decaspermum* Forster & Forster f. has revealed inconsistencies of characters used in generic circumscriptions (Snow 1999, 2000; Snow & Guymer 2001b; J. Dawson, pers. comm.). The following abbreviated key can be used to separate the genera cited by Landrum (1988), the genera revised herein, *Myrtus*, and several related genera. A more detailed key treating only *Lenwebbia*, *Gossia*, and *Austromyrtus* appears below.

Key to the Genera Placed with Myrtus

1. Placenta apical; flowers 4-merous.
    2. Stigma capitate; ovules 2 per locule; placenta simple. *Amomyrtella* Kausel
    2. Stigma not swollen relative to style; ovules many per locule; placenta lamelliform.
        3. Gynoecium 3–4-locular; ovules irregularly attached on placenta; Australia. *Lenwebbia* N. Snow & Guymer
        3. Gynoecium 2–3-locular; ovules uniseriate along margins of placenta; New Zealand. *Lophomyrtus* Burret
1. Placenta subapical or central; flowers 4- or 5-merous.
    4. Embryos mostly circinate. *Gossia* N. Snow & Guymer
    4. Embryos C-shaped.
        5. Bark of branchlets soon peeling in four distinct strips; mature fruits whitish gray with dark spots. *Austromyrtus* (Nied.) Burret
        5. Bark of branchlets not peeling in four distinct strips; mature fruits dark.
            6. Ovary unilocular, barely septate; mature fruits red or yellowish. *Neomyrtus* Burret
            6. Ovary of 2 or more locules; mature fruits dark purple to blackish.
                7. Flowers solitary and nodding.
                    8. Petals not or barely reflexing from anthers at anthesis (except *U. candollei*); bracteoles generally persistent into fruit, ovules in 2–4 rows; seeds 1–2 mm long; scalariform perforation plates present (Schmidt & Baas 1984); America. *Ugni* Turcz.
                    8. Petals mostly reflexing from anthers at anthesis; bracteoles soon deciduous; stigma not swollen to slightly capitate; ovules in 2 rows; scalariform perforation plates absent (Schmidt & Baas 1984); Australasia. *Uromyrtus* Burret
                7. Flowers not solitary or not nodding.
                    9. Stigma capitate to peltate.
                        10. Ovary 4–5- (–8-) locular; ovules 2–3 per locule, the seeds of each locule encapsulated by a moderately thick membranous layer (Snow 1999, 2000). *Decaspermum* Forster & Forster f.
                        10. Ovary 2–3-locular; ovules generally 10 or more per locule, biseriate, the seeds not encapsulated by a membranous layer (Snow 1999, 2000).
                            11. Seeds partially separated horizontally by membranous partitions; Australia, New Caledonia. *Archirhodomyrtus* Burret
                            11. Seeds not partially separated horizontally by membranous partitions; New Caledonia. *Myrtastrum* Burrett
                    9. Stigma narrow, not or barely swollen relative to style.
                        12. Branchlets lacking wings; ovule placentation biseriate, ovules 2–14 per locule; stamens 9–75; anthers with a single oil gland. *Myrteola* O. Berg
                        12. Branchlets winged; ovule placentation multiseriate, ovules 20–40 per locule; stamens 150–220; anthers usually with 3–7 oil glands. *Myrtus* L.

Key to Austromyrtus, Lenwebbia, and Gossia

1. Ovary 3–4-locular; placenta axile, peltate to lamelliform, apically or subapically attached; embryo C-shaped; cotyledons straight, not reflexed, each forming a half-cylinder and together equaling the diameter of the subtending hypocotyl; midrib of adaxial leaf blade often sunken proximally. *Lenwebbia*
1. Ovary mostly 2-locular; placenta axile, slightly capitate (often irregularly so), attached at midpoint or lower; embryo C-shaped or circinate; cotyledons straight to circinate, sometimes reflexed, usually not half-cylindrical in shape; midrib of adaxial leaf blade flush or raised proximally.
   2. Bark of branchlets (growth of current year) soon peeling in four distinct strips; inflorescence usually a solitary flower (="monad" of Briggs and Johnson, 1979) in leaf axils or very rarely a simple dichasium (="botryoid"); peduncle often somewhat reflexed; mature fruits whitish or grayish with small dark spots; embryo C-shaped, endosperm absent; essential oils primarily isobaeckeol. *Austromyrtus*
   2. Bark of branchlets smooth or peeling irregularly, but never in four distinct strips; inflorescence a solitary flower ("monad") or more often a raceme ("botryoid") or panicle ("metabotryoid"); peduncles usually erect; mature fruits mostly dark purple to blackish (rarely red); embryo circinate or C-shaped, endosperm sometimes present; essential oils primarily monoterpenes or sesquiterpene lactones. *Gossia*

SPECIES CONCEPT

De Queiroz (1998, 1999) recently suggested that a general consensus exists regarding most modern definitions of species concepts (see also Cracraft, 2000), in contrast to the abundance of recent literature on the topic that suggests otherwise (Ereshefsky 1992; Claridge et al. 1997; Howard & Berlocher 1998; Wilson 1999; Wheeler & Meier 2000). He calls this general consensus the General Lineage Species Concept (de Queiroz 1988, 1999), since the majority of species concepts accept species as time-extended portions of population lineages. De Queiroz (1998, 1999) further suggests that most disagreements regarding species concepts are actually differences of opinion regarding which methodological approaches, or "species criteria," are most appropriate and most accurate for diagnosing the existence of species lineages. We accept the fundamental distinction between species concepts and species criteria as described by de Queiroz (1998, 1999), as do some other authors (Wiens & Servedio 2000; Snow & Guymer 2001a), and believe it represents a significant move forward in what at times has seemed to be an endless and fruitless discussion.

We used a diagnosability criterion (Davis & Nixon 1992) for species circumscription, which was originally formulated as a phylogenetic species concept (Cracraft 1983; but see particularly Nixon & Wheeler, 1990; Davis & Nixon 1992; Davis 1997, 1999). Diagnosability, using character state data as the empirical basis of species recognition, increasingly is recognized as an inevitable part of the taxonomic process (Davis & Nixon 1992; Snow 1997; Wiens & Servedio 2000; de Pinna 2000). One or more fixed characters or character states were used to delimit species, which are hypothesized to represent hierarchically nested genealogical lineages that exhibit tokogenetic closure (Davis 1997). This is a conservative approach to species recognition that requires character fixation as the minimum empirical evidence necessary to designate specific status to an aggregation of populations (Davis & Nixon 1992) or specimens (Snow 1997, and below), and it accords with the principles of phylogenetic systematics initially set forth by Hennig (1966) (see also Davis 1997, 1999).

Character fixation is highly appropriate as a diagnostic criterion from the perspective of evolutionary pattern (systematics) at the species level, because it removes ambiguity about whether a species can be recognized consistently. Character fixation is also appropriate as a diagnostic tool from the perspective of evolutionary processes (population genetics), because it suggests that gene flow is not occurring between species that are recognized on this basis (Davis & Nixon 1992). Furthermore, apart from the possibility of parallel fixation in populations that are part of separate genealogical lineages, populations that collectively exhibit fixation of at least one character represent members of a species lineage as defined by the General Lineage Concept (de Queiroz 1998, 1999). As with all species criteria (see de Queiroz 1999), the diagnosability criterion may not accurately recover the history of genealogical lineages existing at the current time (Davis & Nixon 1992; Frost & Kluge 1994; Baum & Donoghue 1995); however, this possibility is an insufficient reason to abandon diagnosability, as defined by character fixation, as the preferred methodological tool for species recognition. A diagnosability criterion that requires character fixation as the minimum empirical basis of species recognition meets the theoretical needs of testable and repeatable approaches to species recognition.

When applied consistently, species criteria can and often do affect the number of species recognized by taxonomists (Cracraft 2000). This fact has been demonstrated amply in the ornithological literature (Cracraft 1992; Peterson & Navarro-Sigüenza 1999), where species numbers typically increase under a "phylogenetic" species concept [or, more accurately, a diagnosability criterion] relative to a "biological" species concept [or, more accurately, a reproductive isolation criterion]. In contrast, Snow (1997) suggested that for vascular plants, the use of a species criterion that requires the presence of one or more fixed morphological characters as the minimum evidence needed to confer species status generally will decrease the number of species recognized. This monograph offers additional support for this view. For example, of the approximately twenty putatively "new" species informally recognized in Australia in *Austromyrtus* sensu lato prior to this study (see generic listing of *Austromyrtus* in Henderson, 1997), our consistent implementation of a diagnosability criterion that required one or more fixed morphological characters resulted here in the formal description of only ten new species. Stated another way, many of the putatively "new" species in *Austromyrtus* s. l. listed in Henderson (1997) showed local polymorphic (but nonfixed) variation at the population level. Viewed from the perspective of the evolutionary process, these populations represent variation that typically is expected to occur between populations; however, because these variations were not fixed at the population level, they do not represent a fixed emergent property that can be used as an empirical basis to justify the description of a new species.

It is of interest to note that, although the species criterion of fixed character variation used here differs from approaches to species recognition advocated in earlier times by workers, such as van Steenis (1957) and Bentham and Hooker (Stevens 1997), the approaches converge in their preference of recognizing fewer, often widespread and highly polymorphic species, as opposed to recognizing a larger number of species that include local variants that are not fixed for at least one morphological feature.

DISTRIBUTION

It was necessary to complete the present study of generic boundaries in *Austromyrtus* s.l. before reliable statements could be made about the biogeography of *Myrtus* (Landrum 1988). *Austromyrtus*, as redefined here, and *Lenwebbia* are restricted to Australia. *Austromyrtus* is distributed along the coast and somewhat inland from New South Wales through southeastern Queensland, and *Lenwebbia* occurs from the mountains of northern New South Wales to the mountains of northern Queensland, although its range is interrupted north of the Glasshouse Mountains in southern Queensland. *Gossia* extends from New Guinea east to the Solomon Islands and Vanuatu (=New Hebrides), but has centers of endemism in northeastern New South Wales/southeastern Queensland, northeastern Queensland, and New Caledonia (Snow 2000).

TAXONOMY

NOTE: Abbreviations in specimen citations follow the *Flora of Australia* series (Australian Biological Resources Study 1993), such as: NP (National Park); Ra. (mountain range); R. (River), and more specifically for Queensland, follow the abbreviations of Hyland (1983): SF (State Forest), SFR (State Forest Reserve); LA (Logging Area); EP (Experimental Plots), TR (Timber Reserve), NPR (National Park Reserve). Conservation assessments were determined using the IUCN categories (Species Survival Commission 1994) and Forster et al. (1991). Designations of types of forests follow the system of Webb (1959, 1968) and Webb et al. (1984). See Appendix (p. 89) for "HISPID" names, which are used widely throughout Australian herbaria for tentatively recognized, undescribed taxa (see Henderson, 1997).

Austromyrtus (Niedenzu) Burret, Notizbl. Bot. Gart. Berlin-Dahlem 15: 500. 1941. *Myrtus* sect. *Austromyrtus* Niedenzu, Nat. Pflanzenfam. 3(7): 66. 1898.—LECTOTYPE, designated by McVaugh, 1956b: *Austromyrtus tenuifolia* (Smith) Burret.

Plants suffrutescent to shrubby. Branchlets rounded to laterally compressed, not winged, bark soon peeling in four distinct strips (Fig. 1A), sparsely sericeous to sericeous, oil glands absent or sparse and indistinct. Leaves decussate, opposite or rarely disjunct opposite, evenly distributed along branches, coriaceous, more or less discolorous, matte to slightly glossy above. Stipules obscure or consisting of relatively stout ferrugineous setose hairs. Petioles orangish brown to reddish brown, sometimes channeled slightly above, glandular. Leaf blades pinnately veined, base cuneate, not stem-clasping, apex acute to acuminate, margins flat to slightly revolute, adaxially sparsely sericeous when young but becoming mostly glabrous; oil glands on adaxial surface present but generally obscure (use magnification). Inflorescence a monad or rarely a triad (=simple dichasium), anthopodia and metaxyphylls (Briggs & Johnson 1979) absent. Bracteoles 2, narrowly triangular, mostly rigid, midrib absent, apex usually not extending to the base of calyx lobes, sparsely sericeous, mostly caducous in fruit. Hypanthium cupulate, tube not extending beyond ovary apex, oil glands common, mostly sericeous to villous. Calyx lobes 5, lobes separate in bud, greenish, triangular to broadly ovate, persistent and ascending to erect in fruit, apex obtuse to acute. Petals 5, concave, white or yellowish white; adaxial surface glabrous to sparsely sericeous; abaxial surface shortly villous or sparsely villous, oil glands absent or sparse and faint. Stamens multiseriate, mostly included, folded centerwards in bud; anthers subglobose to globose, dorsifixed, versatile, dehiscing by longitudinal slits, tip of connectives bearing a single gland. Styles 2.5–5 mm long, straight or somewhat apically recurved; stigma terete and not swollen relative to style. Ovary 1, locules 2, placenta axile, irregularly capitate and sometimes only slightly projecting from ovary wall, ovules 6–10 per locule, arranged irregularly. Fruit a fleshy berry, subglobose to globose, base rounded, whitish, grayish, or light green-blue with dark spots at maturity. Seeds irregularly rounded to reniform, laterally compressed, smooth, light brown to ivory or with darker center, micropylar operculum present but sometimes indistinct, adjacent seeds not fused, seed coat hard or bony. Embryo C-shaped, oil glands absent, endosperm absent; hypocotyl relatively thick, tip near radicle not swollen or barely so, held at same plane as embryo; cotyledons straight (not folded), shorter than hypocotyl, folded forwards. Seedling stems terete. Chromosome number unknown.

Austromyrtus, as here circumscribed, is a genus of three species that are restricted to east-central Australia along the coasts to somewhat inland in northern New South Wales and southern Queensland. Its distribution corresponds to the McPherson-Mcleay and Eastern Queensland endemic areas of Crisp et al. (1995, 1999).

Key to the Species of Austromyrtus

1. Leaves linear to narrowly elliptic, 0.15–0.25 (–0.6) cm wide. 3. *A. tenuifolia.*
1. Leaves narrowly elliptic to narrowly ovate, 0.2–1.0 (–1.3) cm wide.
    2. Abaxial leaf surface densely appressed-sericeous. 1. *A. dulcis.*
    2. Abaxial leaf surface glabrous or nearly so. 2. *A. glabra.*

1. Austromyrtus dulcis (C.T. White) L. S. Smith, Proc. Royal Soc. Queensl. 67: 35. 1956.
Myrtus dulcis C.T. White, Proc. Royal Soc. Queensl. 50: 76. 1939.—Type: Australia. Queensland: Moreton District, Palm Beach, between Currumbin and Tallebudgera Creeks, 7 Nov 1929, *C. T. White 6506* (holotype: BRI!; isotype: NY!).

Myrtus tenuifolia var. *latifolia* Maiden & Betche, Census N. S. W. Plants, 143. 1916.—Type: Australia. New South Wales: Lennox Head, Nov 1891, *W. Bauerlen s.n. (NSW 144679)* (lectotype, here designated: NSW!).

Plants 0.2–1.5 m tall, prostrate to spreading shrubs. Bark of trunk stringy or somewhat flaky, gray, brown, or orangish. Branchlets brown to reddish brown or dark gray. Leaf blades 1.0–3.0 (–3.8) cm long, (0.3–) 0.7–1.0 (–1.3) cm wide, narrowly elliptic to narrowly ovate or sometimes slightly falcate, the tip mucronate; adaxial surface sericeous when young but becoming glabrous, oil glands common, midvein flush or slightly raised, matte or slightly glossy; abaxial surface appressed-sericeous and very short-tomentose, oil glands common but obscured by hairs, lateral and intramarginal obscured by hairs. Inflorescences usually axillary, rarely terminal; flowers solitary in leaf axils. Peduncles 5–11 mm long, flexible, often reflexed, sparsely sericeous. Bracteoles 1.5–3.3 mm long, 0.5–1 mm wide. Hypanthium 1.5–2 mm long. Calyx lobes sparsely sericeous to tomentose above, sericeous below, especially on margins. Petals 2.5–4 mm long, 3–4 mm wide, widely ovate to widely obovate. Stamens 50–100; filaments 1.5–5 mm long; staminal disk short-villous; ovary apex densely short hairy. Styles glabrous or sparsely sericeous at base. Berry 4–9 mm long, 3.8–10 mm in diameter, whitish to grayish, glabrous to densely sericeous; seeds 1–2 (–7). Fig. 3.

Phenology. Flowering mostly October through January (–May); fruiting January through June.

Distribution (Fig. 4). From northern New South Wales to southeastern Queensland; along the coasts in forests on hillsides or sandy soils; sea level to 140 m.

Vernacular name. Midgen berry.

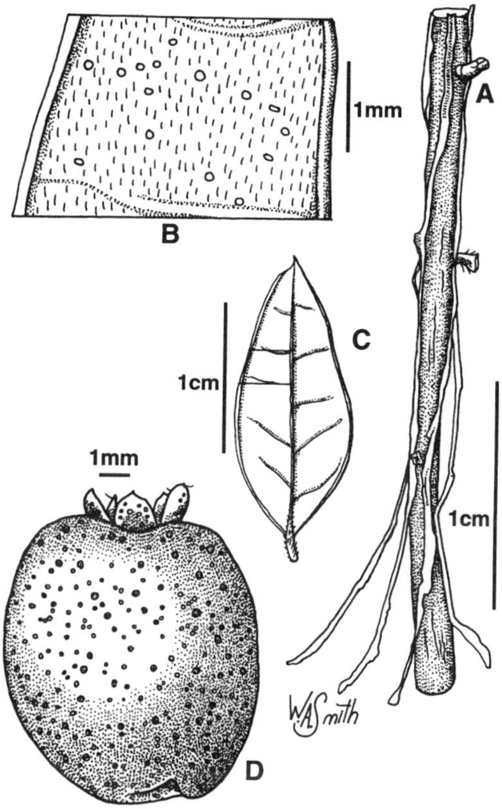

FIG. 3. *Austromyrtus dulcis*. A. Older branch with peeling strips of bark. B. Detail of abaxial leaf surface, showing densely appressed pubescence. C. Leaf, abaxial view. D. Mature fruit with dark spots. (Based on: A, C, *Everist 7752*; B, *Sharpe 4312*.)

REPRESENTATIVE SPECIMENS. **Australia.** QUEENSLAND. WIDE BAY DISTRICT: Fraser Is., on dunes above swamp near Eurong Beach, *Baxter 774* (BRI, CANB); Fraser Is. on dune bordering swamp between hairpin dunes S of Erong Beach, 25°34′S, 153°06′E, *Baxter 776* (BRI); Fraser Is., Sandy Cape, *Blake 22696* (BRI); 1 km W of Poona Point, 25°43′S, 152°54′E, *Forster PIF13915 & Smyrell* (BRI); Cootharaba, Dec 1919, *Francis s.n.* (BRI 093398); Cooloola, Laguna Bay, southern extremity of Cooloola SF, 26°16′S, 153°03′E, *Harrold 290* (BRI); Noosa NP Noosa, ca. 1 km S of car park, 26°23′S, 153°05′E, *Sharpe 4283* (BRI); Noosa NP, Noosa, ca. 800 m N of Park Edge Rd, 26°23′S, 153°05′E, *Sharpe 4288* (BRI); Fraser Is., in wallum adjacent to vine forest, *Webb & Tracey 6373* (BRI); Fraser Is., dunes N of Middle Rock, S of Orchid Village, *Whaite & Whaite 3495* (NSW); Noosa Heads, Jan 1920, *White s.n.* (BRI 093399).—MORETON DISTRICT: Bribie Is., top end off Deepwater Point, 26°49′S, 153°08′E, *Batianoff 1368 & Elsol* (BRI); Stradbroke Is., *Blake 7092* (BRI); Brisbane, Dec 1901, *Boorman s.n.* (NSW 144684); Moreton Is., ca. 4.5 km NE of Tangalooma, 27°05′S, 153°24′E, *Durrington 206* (BRI); North Stradbroke Is., Eighteen Mile Swamp, 27°37′S, 153°28′E, *Durrington 676* (BRI); Moreton Is., 17 km SSE of Tangaboma, 27°20′S, 153°25′E, *Durrington 1197* (BRI); Marcus Beach, 26°33′S, 153°03′E, *Forster PIF15237* (BRI); Mudjimba Beach, 22 km N of Caloundra, 26°38′S, 153°06′E, *Franks AJF9209009 & Suosaari* (BRI); Southport, *Gordon 118* (BRI); Fig Tree Pocket, Brisbane, 27°28′S, 153°01′E, *Guymer 1991B* (BRI); Noosa NP, 26°24′S, 153°04′E, *Guymer 2128A* (BRI); Stradbroke Is., between Amity Point and Point Lookout, *Hubbard 2298* (BRI); Burleigh Head, *Michael 2256* (BRI, QRS); 3 km N of Coolum Beach, ca. 130 km N of Brisbane, *Sharpe 2028* (BRI); Mt Coolum, ca. 3 km S of Coolum Beach, 26°33′S, 153°04′E, *Sharpe 3127* (BRI); Mt Coolum, ca. 3 km S of Coolum Beach, 26°33′S, 153°05′E, *Sharpe 4287* (BRI); Pimpana Is., *Shirley s.n.* (BRI 093413); Burleigh Heads NP, trail along South entrance, *Snow & Bradford 7457* (BRI, CANB, GREE); Nerang, Dec 1913, *White s.n.* (BRI); Stradbroke Is., Apr 1917, *White s.n.* (BRI 093423); Bribie Is., Feb 1919, *White s.n.* (BRI 093415); Burleigh Heads, Nov 1921, *White s.n.* (BRI).—NEW SOUTH WALES: Chinaman's Beach, just S of Evans Head, 29°07′S, 153°26′E, *Bean 8010* (BRI); Cape Byron, Mar 1896, *Betche s.n.* (NSW 144675); Shelley Beach Head, S of Yamba, *Blaxell 132* (NSW); Byron Bay, Sep 1916, *Cheel s.n.* (BRI, NSW 144677); Chinaman's Beach ca. 2 mi (3.2 km) S of Evans Head, *Coveny 4638* (BRI, MO, NSW); southern outskirts of Brunswick Heads, 28°32′S, 153°33′E, *Dunn et al. 112* (BRI, MO); Hungry Head, Apr 1980, *Floyd 1471* (NSW); 500 m S of Oyster Ck entrance, 3 km N of Valla, *Floyd 1672* (NSW); New Brighton, near Billinudgel, Jan 1936, *Fraser s.n.* (NSW 144674); Hungry Head, Urunga, 30°31′S, 153°02′E, *Guymer 297* (BRI); Tweed Heads, *Kenny s.n.* (BRI AQ093396); Byron Bay, Nov 1903, *Maiden & Boorman s.n.* (NSW 144676); Narrowneck, *Smith 5104* (BRI, RSA); Mooball Beach, 28°25′S, 153°40′E, *Stocker 1460* (QRS); 100 m inland from Black Rock camping area, Bunjalung NP, 29°06′S, 153°23′E, *Wiecek 408 & Porteners* (BRI, MEL).

Austromyrtus dulcis is the most common and wide-ranging species in the genus, and it is sometimes cultivated as a ground cover along the edges of footpaths. The grayish fruits with dark spots easily separate *A. dulcis* and its congeners from other genera of

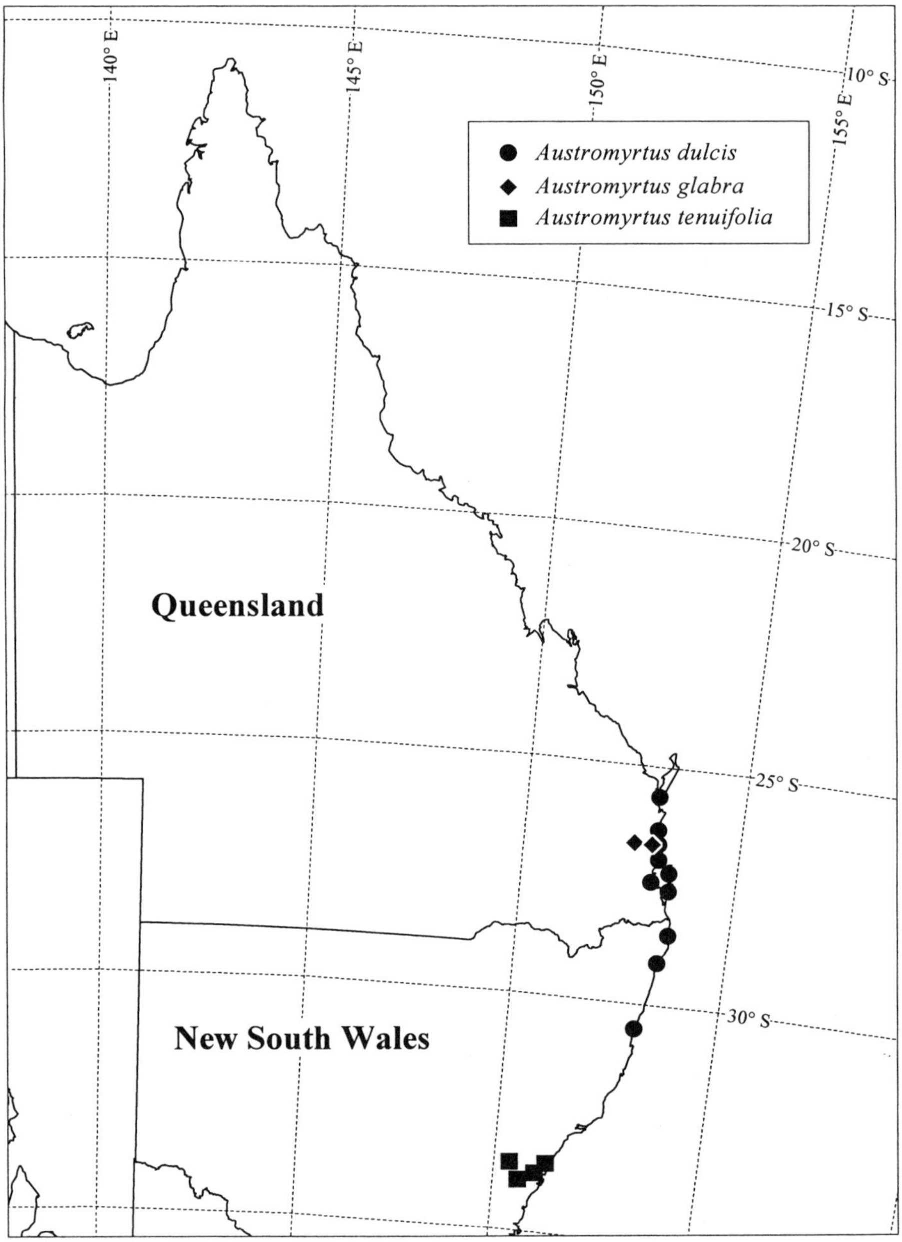

FIG. 4. Distribution of *Austromyrtus dulcis*, *A. glabra*, and *A. tenuifolia*.

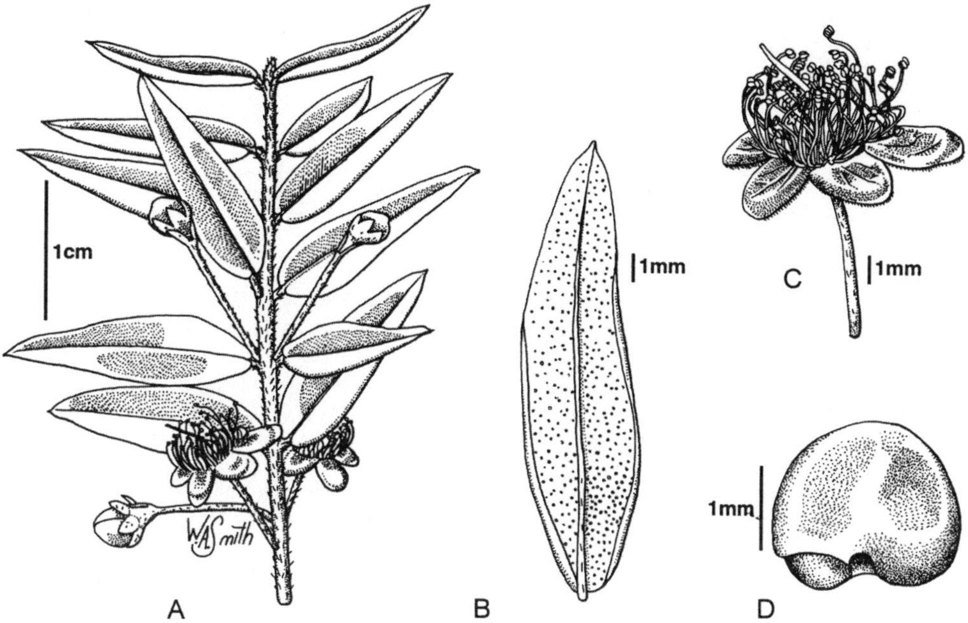

FIG. 5. *Austromyrtus glabra*. A. Flowering branch. B. Leaf, abaxial view. C. Flower. D. Seed; arrow indicates micropylar operculum. (Based on: A, B, *Bean 2681*; C, *Guymer 2186*; D, *Forster 15057*.)

Myrtaceae. Although *A. tenuifolia* is also in New South Wales, its range is more inland and at higher elevations.

The specific epithet derives from Latin *dulcis*, in reference to the sweet fruits, which have a distinctive, cinnamon-like taste.

2. Austromyrtus glabra N. Snow & Guymer, sp. nov.—TYPE: AUSTRALIA. Queensland: Moreton District, about 3.5 km NW of Cooloolabin, NW of Yandina, 152°52′E, 26°32′S, 23 Nov 1987, *G. P. Guymer 2186* (holotype: BRI!; isotypes: A! AD! B! BRI! CANB! DNA! GREE! JEPS! K! L! MEL! MO! NE! NSW! NY! PERTH! QRS! SING! UPS! US!).

Ad instar *Austromyrtus dulcis* autem folia matura pagina inferiore glabra vel fere ita.

Plants 0.3–1.3 m tall, prostrate to spreading shrubs. Bark of trunk smooth to stringy, gray or orangish to reddish brown. Branchlets light brown to orangish or reddish brown. Leaf blades 0.8–3.2 cm long, 0.2–1.0 cm wide, narrowly elliptic to narrowly ovate or rarely falcate, the tip acuminate to mucronate; adaxial surface glabrous, oil glands abundant, midvein flush, matte or slightly glossy; abaxial surface glabrous or very sparsely shortly sericeous, oil glands common to abundant, lateral and intramarginal veins indistinct or invisible. Inflorescence axillary or terminal, flowers solitary or rarely in fascicles of 2 or 3. Peduncles 1.5–9 mm long, rigid, glabrous to sericeous. Bracteoles 1.1–2.5 mm long, 0.5–0.8 mm wide. Hypanthium 1.3–1.7 mm long. Calyx lobes sparsely to densely short-tomentose above, sparsely sericeous below. Petals 2–4 mm long, 2.5–4 mm wide, widely elliptic to oblate. Stamens 85–95; filaments 2–4 mm long; staminal disk glabrous to short-villous. Styles glabrous; ovary apex shortly hairy. Berry 3.5–5.5 mm long, 3.5–5.5 mm in diameter, white to gray with dark spots, glabrous to minutely glabrescent; seeds 2–5. Fig. 5.

Phenology. Flowering in November and December; fruiting from March through April.

Distribution (Fig. 4). Southeastern Queensland; occasional to locally common in open eucalypt forests, wet sclerophyll forests, or wallum (=relatively low vegetation in coastal acidic sandy soils in southeastern Queensland); 15–380 m.

REPRESENTATIVE SPECIMENS. **Australia.** QUEENSLAND. WIDE BAY DISTRICT: Oakey Ck Rd, Mapleton SF, 26°36′S, 152°51′E, *Bean 2681* (BRI, CANB); SF 351, 4 km SE of Eumundi, 26°30′S, 152°59′E, *Forster PIF15057* (BRI, MEL); SF 1239 Kenilworth, Mapleton forest drive, 3 km NW of Mapleton, 26°37′S, 152°51′E, *Forster PIF15179* (BRI); ca. 8 km N of Mapleton, 26°35′S, 152°25′E, *Guymer 2129* (BRI, MO); Como SF, Scientific Area No. 3, *Sandercoe C1295* (BRI); Caravan Park, 18 km N of Maroochydore on Noosa Rd, 26°30′S, 153°06′E, *Sharpe 180* (BRI); Blackall Ra. ca. 3.5 km NW of Coololabin on forestry track to Mt Glorious lookout, 26°32′S, 152°52′E, *Sharpe 4312* (BRI); Blackall Ra. ca. 3.5 km NW of Coololabin on forestry track to Mt Glorious lookout, 26°32′S, 152°52′E, *Sharpe 4314* (BRI); 1 km N of Peregian Eumundi, Nov 1892, *Simmonds s.n.* (BRI 093421); Kiamba, *Thorne et al. 21208* (RSA).—MORETON DISTRICT: SF 351, 4 km SE of Eumundi, 26°30′S, 152°59′E, *Bean 7077* (BRI); Caloundra, *Clemens 42509* (RSA); Kureelpa Falls Rd, Kureelpa, ca. 8 km NW of Nambour, 26°36′S, 152°53′E, *Sharpe 4924 & Bean* (BRI, MEL); Ferntree Ck N.P., Kulangoor, ca. 5 km N of Nambour, 26°37′S, 152°58′E, *Sharpe 4836 & Chapman* (BRI).

Austromyrtus glabra is easily distinguished from *A. dulcis* by the glabrous abaxial leaf surface. Their ranges overlap in the vicinity of Caloundra, and Stradbroke and Fraser islands. The species maintain their differences in abaxial leaf pubescence in sympatry, strengthening our view that *A. glabra* should be recognized as a distinct species.

The specific epithet, the Latin *glabra,* is in reference to the glabrous undersurface of the leaves.

3. Austromyrtus tenuifolia (Smith) Burret, Notizbl. Bot. Gart. Berlin-Dahlem 15: 501. 1941. *Myrtus tenuifolia* Smith, Transact. Linn. Soc. 3: 280. 1797.—TYPE: AUSTRALIA. New South Wales: Port Jackson, 1795, *W. White s.n.* (holotype: BM! LINN-SMITH 882.3, photocopies: BRI!).

Plants 0.7–2.5 m tall, erect, often multi-stemmed shrubs. Bark of trunk smooth to scaly or tessellated, brown to gray. Branchlets brown to reddish brown. Leaf blades (1.2–) 2.0–3.0 (–4.5) cm long, 0.15–0.25 (–0.6) cm wide, linear or rarely narrowly elliptic, the tip pointed to mucronate; adaxial surface glabrous or glabrescent, oil glands common, midvein flush or slightly impressed, matte; abaxial surface sericeous but becoming glabrous, oil glands common, lateral and intramarginal absent or faint. Inflorescence axillary, flowers solitary. Peduncles 7–10 mm long, rigid, sparsely sericeous to villous. Bracteoles 1–2.5 (–4) mm long, 0.3–0.8 mm wide. Hypanthium 0.5–1.7 mm long. Calyx lobes sparsely short-villous to glabrescent above, sericeous to villous below. Petals 3–4 mm long, 2.4–3.5 mm wide, widely elliptic to widely ovate. Stamens up to 75; filaments 2–4 mm long; staminal disk sparsely glabrescent. Styles glabrous; ovary apex glabrous. Berry 5–8 mm long, 3.5–5 mm in diameter, grayish to greenish blue, sparsely sericeous; seeds 4–13.

Phenology. Flowering November through January; fruiting January and February.

Distribution (Fig. 4). New South Wales; locally common in wet sclerophyll forests, sandstone and rocky areas, and especially along streambanks and floodplains; 60–670 m.

ADDITIONAL SPECIMENS EXAMINED. **Australia**. NEW SOUTH WALES: Mossman's Bay, *Betche s.n.* (NSW 144653); Spring Ck, Hornsby, Nov 1914, *Blakely s.n.* (NSW 144670); Berowra, *Boorman s.n.* (BRI 093542); Waonora R., Nov 1898, *Boorman s.n.* (NSW 144645); Sydney Middle Harbour, Port Jackson, 1802–5, *Brown s.n.* (CANB, MEL 238768); Middle Harbour, Sydney (Port Jackson), 1803, *Brown s.n.* (NSW 180701); Sassafrass Gully, Springwood, Dec 1894, *Camfield s.n.* (NSW 144658); Galston Gorge, *Burgess 11046* (CANB); Blacksmith's Ck, ca. 5 mi N of Bilpin, Nov. 1958, *Constable s.n.* (NSW 48222); Pennant Hills Park, Cheltenham, 33°45′S, 151°05′E, *Coveny et al. 12101* (BRI, MO, NSW); Devlin's Ck in Pennant Hills Park, Cheltenham, 33°45′S, 151°05′E, *Coveny 11438 & Dunn* (CANB, MO, NSW, RSA); Blue Mts, 33°36′S, 150°15′E, *Cunningham 49* (CANB, NSW); Upper Lane Cove R., Mar 1929, *Fletcher s.n.* (NSW 144664); Hawkesbury, *Deane s.n.* (NSW 144650); Pennant Hills, *Fraser s.n.* (NSW 144671); Devlins Ck, Pennant Hills Park, 33°44′S, 151°04′E, *Goldsack 719, 720, 721* (BRI); Springwood, Nov 1913, *Hamilton s.n.* (NSW 144656); Colo Gorge below Boorai Ridge, 33°15′S, 152°32′E, *Hind 898* (NSW); Bowens Ck, a little upstream from road bridge, 33°29′S, 150°28′E, *Hind 2203* (ASU, BRI, NSW); Bowens Ck crossing of Mt Wilson–Bilpin road, 15.9 km from Mt Wilson post office, 33°30′S, 150°29′E, *Hind 5565* (NSW); Junction of Cabbage Tree Ck and Burralow Ck, ca. 9 km SW of North Richmond, Vale of Avoca Rec. Reserve, 33°36′50″S, 150°37′40″E, *James 1219 & Stanberg* (BRI); Kangaroo Ck, gully near Ck, Jan 1946, *Johnson s.n.* (NSW 144646); above Flat Rock Ck, Jun 1965, *Johnson s.n.* (NSW 144652); Tascott, Nov 1953, *King s.n.* (NSW 144646); Lane Cove, Nov 1937, *Lemberg s.n.* (NSW 144650); Port Jackson District, *Maiden s.n.* (NSW 14655); Berowra Railway Station, to a gully leading into Cowan Ck, Dec 1887, *Maiden s.n.* (NSW 144667); 1 mi E of Cowan, *McBarron 8951* (NSW); Woy Woy Ck, Woy Woy, 33°30′S, 151°19′E, *McBarron 17521* (NSW); Northbridge, *McKee 566* (MEL, US); Rouse Hill, upper Cattai Ck, 33°41′S, 150°55′E, Jun 1989, *Mills s.n.* (NSW 251556); Black Rocks, Parramatta, Jan 1845, *Mueller s.n.* (NSW 144654); McCarrs Ck below road crossing, just above waterfall, *Rodd & Watson 3519* (NSW); on banks of Nepean R., ca. 4 mi S of Wallacea, *Salasoo 2686* (NSW); Bent's Basin, Nepean R., 7 km S of Wallacia, 33°56′S, 150°38′E, *Telford 7221* (BISH, CANB); Galston Gorge near Sydney, *White 10340* (BRI); Pennant Hills Park, Cheltenham, *Wilson et al. 129* (BRI, NSW); Middle Ck, at the foot of Oxford Falls, Oct 1982, *Wilson s.n.* (UNSW 13677) (NSW, QRS); Bents Basin, *Woolls s.n.* (MEL 238766); St. Ives, on creek bank below Tree Fern Gully, Ku-Ring-Gai Wildflower Garden, *Wrigley* 22288 (CANB).

The linear leaves of *A. tenuifolia* distinguish the species from *A. dulcis* and *A. glabra*; they are somewhat aromatic when fresh. This species is generally more erect and less sprawling than its congeners, and its fruit often has somewhat of a greenish tinge.

Austromyrtus tenuifolia is often cultivated as an ornamental. The specific epithet derives from the Latin *tenuis* (narrow) and *folium* (leaf) in reference to the narrow leaves.

Lenwebbia N. Snow & Guymer, gen. nov.—TYPE: *Lenwebbia lasioclada* (F. Mueller) N. Snow & Guymer.

Arbores vel frutices, *Austromyrtum* sensu stricto simulantes autem ovarium 3–4-loculare placenta peltata vel lamelliformi apicali (vel subapicali) instructum. A *Lophomyrto* ovulis non in seri sola circum marginem placentae differt.

Trees or treelike shrubs. Bark of trunk smooth, brown to gray. Branchlets rounded to laterally compressed, not winged, brown, smooth; oil glands common but obscured by hairs. Leaves discolorous, matte above and below, decussate, opposite, evenly distributed along branches, coriaceous. Stipules consisting of relatively stout ferrugineous setose hairs. Petioles 3–9 mm long, olive green to brownish, glandular or eglandular. Leaf blades pinnate, narrowly elliptic to elliptic or occasionally obovate, base cuneate, not clasping the stem, apex mostly acute to slightly acuminate, the tip rounded to pointed, margins flat or often slightly revolute; oil glands of adaxial and abaxial surface visible only with magnification, common to abundant. Inflorescence terminal or axillary, a monad or triad, anthopodia and metaxyphylls absent. Peduncle rigid. Bracteoles 2 and often falling early, more or less rigid, the tips not reaching base of calyx lobes, caducous in fruit. Hypanthium tube not extending beyond ovary apex, with oil glands common to abundant. Calyx lobes

4, fused proximally; lobes separate in bud, persistent in fruit, mostly ascending and held above body of fruit. Petals 4, concave, white to occasionally pinkish; adaxial surface glabrous but somewhat ciliate on margins; abaxial surface glabrous, with oil glands common to abundant, sometimes pronounced and wart-like on upper surface. Stamens multiseriate, exserted, folded centerwards in bud. Staminal disk densely short-hairy; anthers globose to subcylindrical, dorsifixed or basifixed, versatile, dehiscing via longitudinal slits, the tips of connectives with a single gland. Style mostly straight. Fruit a soft berry, dark purplish black, subglobose to globose, the base rounded. Ovary 1; locules 3–4; placenta axile, distinctly peltate to lamelliform, attached apically or subapically; ovules placed irregularly. Seeds irregularly rounded to quite angular, smooth, light brown, center softer, dark brown to blackish; micropylar operculum present but often obscure; seed coat very hard, bony, adjacent seeds separate. Embryo C-shaped, oil glands present but very faint, starchy, endosperm absent; hypocotyl relatively thick, not or barely swollen near tip, held at same plane as embryo; cotyledons straight, shorter than hypocotyl, folded forward. Seedling stems round to flattened. Chromosome number unknown.

The 4-merous flowers with a hairy staminal disc distinguish *Lenwebbia* from other genera of Myrtaceae in eastern Australia.

Lenwebbia is endemic to Australia in northeastern New South Wales and eastern Queensland. The genus has a disjunct distribution. Both species occur in the McPherson-Macleay region in southern Queensland and *L. lasioclada* also in the Atherton region, an area of high endemism in northeastern Queensland (Crisp et al. 1995, 1999).

The generic name honors the eminent plant ecologist Leonard J. Webb (b. 1920). Webb has contributed greatly to the understanding of Australian plant ecology, particularly that of the coastal rainforests in which his eponymous genus resides (Webb 1959, 1968; Webb et al. 1984). He has also collected large numbers of specimens in Myrtaceae.

Key to the Species of Lenwebbia

1. Secondary veins on abaxial leaf surface flush or only slightly raised, often hairy, the hairs whitish. 4. *L. lasioclada*.
1. Secondary veins on abaxial leaf surface prominently raised, usually hairy, the hairs often ferrugineous. 5. *L. prominens*.

4. Lenwebbia lasioclada (F. Mueller) N. Snow & Guymer, comb. nov. *Myrtus lasioclada* F. Mueller, Fragm. 9: 148. 1875. *Austromyrtus lasioclada* (F. Mueller) L. S. Smith, Proc. Roy. Soc. Queensland 48: 35. 1956.—Type: Australia. Queensland: Cook District, Mount Bellenden Ker, *W. Hill 187* (holotype: MEL!; isotype: BRI [fragment]!).

Trees 2–6 (–10) m tall. Branchlets sericeous to densely sericeous. Petioles channeled above. Leaf blades 3.5–12.0 cm long, 1.7–4.5 cm wide, narrowly elliptic to elliptic; adaxial surface glabrous or sericeous along midvein, the midvein impressed near base or often raised; abaxial surface glabrous and often sericeous on midvein and between secondary nerves, with oil glands few to common, secondary veins indistinct to prominent, intramarginal veins absent to prominent, the inner intramarginal vein 1–2 mm from edge. Inflorescence axillary, flowers solitary. Peduncles 10–42 mm long, sparsely sericeous to sericeous or villous. Bracteoles up to 5 mm long, ca. 1 mm wide, narrowly triangular, foliaceous but midrib absent, sparsely sericeous. Hypanthium 3–5.2 mm long, cupulate, sericeous to densely sericeous. Calyx lobes 1.3–3 mm long, broadly triangular, the apex

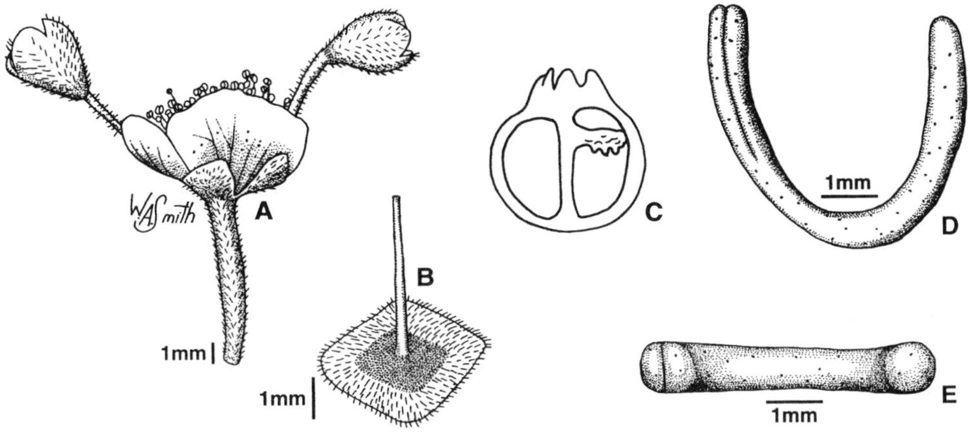

FIG. 6. *Lenwebbia lasioclada.* A.Dichasial inflorescence. B. Ovary apex with hairy staminal ring and style. C. Longitudinal section through fruit (seeds removed), showing subapical protruding placenta in one locule. D, E. Two views of an embryo with appressed cotyledons. (Based on: A, B, *Blake 19123*; C–E, *Guymer 2023*.)

obtuse, glabrous to sparsely sericeous above, glabrous to sometimes densely sericeous on margins below. Petals 5–6 mm long, 4–6 mm wide, widely ovate to widely obovate. Stamens 55–105, filaments 2–6 mm long. Styles 3.5–7.5 mm long, glabrous to sparsely sericeous throughout; ovary apex glabrous; ovules up to 25 per placenta. Berry 9.5–19 mm long, 8.5–19 mm in diameter, sparsely sericeous; seeds in mature fruit 2–28. Embryo with oil glands present but very faint, starchy, endosperm absent; hypocotyl relatively thick, tip near radicle not or barely swollen, at same plane as embryo; cotyledons straight, shorter than hypocotyl, folded forward. Fig. 6.

Phenology. Flowering from December (lowland populations) through June (higher elevations); fruiting mostly January through March.

Distribution (Fig. 7). Northern New South Wales and southern Queensland, and disjunct in northeastern Queensland; rare to locally common in rainforests or exposed sites above 1000 m or in swamps below 100 m.

Vernacular name. Velvet myrtle.

ADDITIONAL SPECIMENS EXAMINED. **Australia.** QUEENSLAND. COOK DISTRICT: SFR 185, Danbulla, Tinaroo LA, 7.5 km past western boundary of Forestry grid, 17°06′S, 145°34′E, *Forster PIF14348* (BRI, MEL); SFR 185, Danbulla, Kauri LA, top of Kauri Ck Rd, 17°05′S, 145°35′E, *Forster PIF14723* (BRI); North Bell Peak, summit area, Malbon Thompson Ra., 17°05′S, 145°52′E, *Forster PIF18039 et al.* (BRI); Longlands Gap SF, 17°26′S, 145°28′E, *Forster PIF21877 et al.* (BRI); SFR 185, Haig LA, 17°06′S, 145°35′E, *Gray 362* (BISH, BRI, CANB, MEL, MO, NSW, RSA); SFR 310, Upper Goldsborough LA, 17°16′S, 145°51′E, *Gray 926* (BRI, CANB, QRS); SFR 607, Emerald LA, 17°05′S, 145°35′E, *Gray 1373* (BISH, BRI, CANB, MEL, MO, NSW, RSA); SFR 194, Parish of East Barron, 17°21′S, 145°27′E, *Gray 2481* (BRI, QRS [seedlings]); 30 km along Mt Lewis Rd, off Julatten Rd, 16°29′S, 145°15′E, *Guymer 2023* (BISH, BRI, MO); Bellenden Ker, 17°15′S, 145°51′E, *Hunter JH1531* (BRI); Bellenden Ker, 17°15′S, 145°50′E, *Hyland 6578* (BRI, CANB, QRS); Bellenden Ker Ra., 17°15′S, 145°50′E, *Hyland 7774* (BRI, QRS); SFR 310 Bellenden Ker Ra., 17°16′S, 145°51′E, *Hyland 10211* (QRS); Mt Bartle Frere, 1892, *Johnson s.n.* (MEL 221564, BRI AQ462803); Mt Bartle Frere, 1891, *Johnson s.n.* (MEL 726890); Russell R., 1892, *Johnson s.n.* (MEL 726893); SFR 185, ca. 5 km N of Tinaroo Falls Dam, 17°05′S, 145°36′E, *Moriarty 1979* (BRI, CANB, NSW, QRS [seedlings]); SFR 185, Danbulla, Emerald LA, 17°09′S, 145°38′E, *O'Farrell 100* (BRI, QRS); Mt Haig, 17°07′S, 145°35′E, *Sankowsky 846 & Sankowsky* (BRI); Topaz, 17°24′S, 145°29′E, *Sankowsky 1355 & Sankowsky* (BRI); Mt Bellenden Ker, 1887, *Sayer 34* (MEL); Mt Bellenden Ker, 1887, *Sayer 80* (MEL); SFR 99, 17°S, 145°E, *Smith 10137* (BRI, CANB,

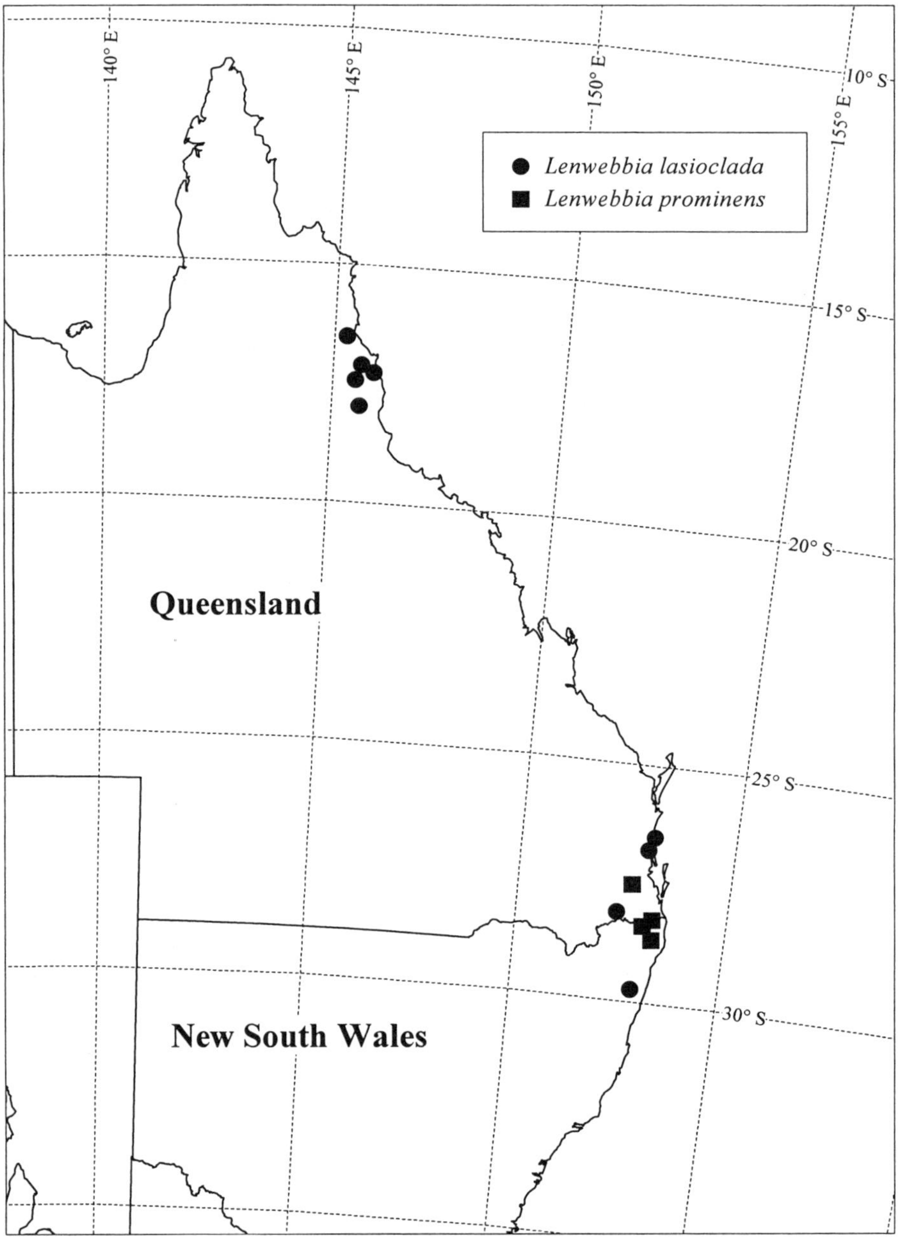

FIG. 7. Distribution of *Lenwebbia lasioclada* and *L. prominens*.

NSW); NW peak of Mt Bartle Frere, 17°23′S, 145°47′E, *Tucker MCT122* (BRI).—NORTH KENNEDY DISTRICT: Herberton Ra., *Kajewski 1382* (BRI, NSW, NY, SING); Walters Waterhole, Koolmoon Ck, S of Ravenshoe, 17°45′S, 145°33′E, *Young 1748* (BRI).—MORETON DISTRICT: Tania Ave, Eudlo Ck Nat. Park, N of Eudlo, 26°42′S, 152°58′E, *Bean 6097* (BRI, NSW); Doonan Ck swamp, 5 km W of Peregian, 26°30′S, 153°03′E, *Bean 7113* (BRI, CANB, MEL); Bruce Hwy, 1 km N of Eudlo Ck, 26°40′S, 153°00′E, *Bean 7147* (BRI); Winston Rd, Palmwoods, 7 km SE of Nambour, 26°41′S, 152°59′E, *Forster PIF15085* (BRI, K, MEL, NSW, QRS); Wilsons Peak, upper slopes, 28°14′S, 152°29′E, *Forster PIF15694 & Leiper* (BRI, NSW, QRS); Mt Roberts ridge, Main Ra., SW of Boonah, 28°13′S, 152°28′E, Feb 1986, *Grimshaw s.n.* (BRI AQ450409); property of Mr Howling, The Wiñery, Mons, ca. 9 km SE of Nambour, 26°41′S, 153°00′E, *Sharpe 4988 et al.* (BRI, K); The Winery, ca. 5 km NW of Buderim on Bruce Hwy, 26°41′S, 153°01′E, *Sharpe 5387* (BRI, NSW); Didillibah Rd, Didillibah, ca. 4 km SE of Nambour, 29°39′S, 152°58′E, *Sharpe 5395* (BRI, MEL, NSW); Winston Rd, Palmwoods, 7 km SE of Nambour, 26°41′S, 152°59′E, *Sharpe 5399 et al.* (BRI, K, MEL, NSW); Mt Cordeaux, Cunningham Gap, ca. 50 km ESE of Warwick, 28°04′S, 152°21′E, *Sharpe 4877 & Guymer* (BRI); Mistake Mountain, *Smith & Webb 3691* (BRI, CANB, K, NY); near Moonshine Valley, Forest Glen turnoff from Bruce Hwy, S of Nambour, 26°41′S, 153°00′E, *Thomas 341 & Barry* (BRI); St. Johns Wood, suburb of Brisbane [from natural population at Palmwoods], *Thompson MOR475A* (BRI).

The setose, ferrugineous stipular hairs of *L. lasioclada* are quite evident. The secondary nerves on the underside of the leaf may be visible or not, but are neither prominent nor raised, as in *L. prominens*. The vestiture of the hypanthium can vary from sparse to dense (e.g., *Bean 7147*,BRI; *Leiper s.n.*, BRI AQ584408), a pattern that can also occur in *L. prominens* (*Forster PIF14614 & Leiper*, A, BRI, CANB, GREE, K, L, MEL, MO, QRS; *Johnson s.n.*, NSW 17303).

Some of the lower altitude populations were designated tentatively as "*Austromyrtus* sp. Blackall Range, *P.R. Sharpe 5387*" (Henderson 1997: 125); however, these populations cannot be separated consistently from upland populations based on our species criterion of presence of fixed morphological differences.

A gap of some 1300 kilometers separates the southern populations of *L. lasioclada* near Nambour from populations in the North Kennedy and Cook districts in northeastern Queensland. Disjunct distributions of this general pattern occur elsewhere in Myrtaceae, such as *Gossia hillii* (no. 11), *Babingtonia angusta* A. R. Bean (Bean 1999), and *Eucalyptus grandis* W. Hill ex Maiden (Chippendale 1988).

Lenwebbia lasioclada has an interesting ecological distribution. Throughout most of its range it has a clear preference for wet, often rocky sites above 1000 meters; however, it also occurs at elevations below 100 meters in wet and swampy sites near Nambour, Queensland. The common ecological denominator seems to be high levels of moisture, which in its range mostly occur at these disjunct elevational zones. A similar ecological disjunction in southeastern Queensland between mesic lowland and upland populations occurs in some species of *Rapanea* in Myrsinaceae (P. Sharpe, pers. comm.). The widespread elevational range of *Babingtonia angusta* from nearly sea level to ca. 1000 meters is particularly similar to that of *Lenwebbia lasioclada*.

Populations of *L. lasioclada* from higher, more exposed elevations in the Cook and Moreton districts have somewhat smaller leaves and more shrublike growth habits, as represented by *Sayer 80* (BRI, MEL), *Hyland 7774* (BRI, QRS) from Mt Bellenden Ker, and by *Forster PIF15694 & Leiper* (BRI, NSW, QRS) from the upper slopes of Wilson's Peak from the Main Range near the NSW-QLD border. The fruits of northern populations tend to be somewhat smaller than those from southern populations.

Regent bowerbirds have been observed eating fruits from a specimen in Lamington NP (*Olsen s.n.*, BRI AQ398234).

The specific epithet is from the Greek *lasios* (woolly or hairy) and *clados* (branch), in reference to the branchlets.

5. Lenwebbia prominens N. Snow & Guymer, sp. nov.—TYPE: AUSTRALIA. Queensland: Moreton District, Upper Mudgeeraba Creek, 30°S, 153°E, *N. Byrnes 4069 & L. W. Jessup* (holotype: BRI!; isotypes: BRI! CANB! L! K! MO! NSW! QRS).

Simulans *L. lasiocladum* autem folia inferiora venis secundariis prominentibus elevatis et saepe pilis ferrugineis ornatis.

Trees 2–10 m tall, the branches occasionally spreading widely. Branchlets sericeous to villous. Petioles sometimes channeled above. Leaf blades 2.5–10.5 (–12.5) cm long, 1.2–4.0 (–4.7) cm wide, elliptic or occasionally narrowly elliptic to obovate; adaxial surface sericeous but becoming glabrous, the midvein flush or impressed; abaxial surface ferrugineous-hirsute or -pilose, with oil glands absent to common, secondary veins generally quite prominent, intramarginal veins prominent, the inner intramarginal vein 0.5–2 mm from edge. Inflorescences axillary, flowers solitary or paired. Peduncles (5–) 9–30 mm long, shortly sericeous to villous. Bracteoles 3–7 mm long, 0.5–1 mm wide, linear, scale-like, with midrib absent or faint, villous. Hypanthium 2.8–5.2 mm long, obconic to cupulate, sparsely sericeous to densely sericeous or densely villous. Calyx lobes 1.3–3 mm long, broadly triangular to broadly ovate, the apex obtuse to acute, sericeous near margins and apex above, sparsely villous to sparsely tomentose below. Petals 2–6 mm long, 4–7 mm wide, ovate to widely obovate or widely elliptic. Stamens 75–275, the filaments 3–7 mm long. Styles 3.5–7 mm long, glabrous to sparsely sericeous near base; ovary apex glabrous or slightly sericeous; ovules up to 16 per placenta. Berry (6–) 9–19 mm long, (6–) 7.5–19 mm in diameter, sparsely sericeous to sparsely villous; 1 seed in mature fruit. Embryo not seen. Fig. 8.

Phenology. Flowering from November through February; fruiting January through September.

Distribution (Fig. 7). Northern New South Wales and southern Queensland; rare to locally common in seasonally saturated sites along banks of streams in rainforest or notophyll vine forests; soils often volcanic and derived from rhyolites; 30–1000 m.

Conservation status. The species is Vulnerable (criterion C2a, Species Survival Commission 1994).

ADDITIONAL SPECIMENS EXAMINED. **Australia.** QUEENSLAND. MORETON DISTRICT: top of Bahnamboola Falls, Lamington NP, 28°S, 153°E, *Blake 19123* (BISH, BRI, MO, NSW); Lamington NP, Mar 1953, *Cribb s.n.* (BRI AQ468549); Lamington NP, Jun 1953, *Cribb s.n.* (BRI AQ468550); Springbrook NP, Blackfellows Ck, 28°12′S, 153°15′E, *Forster PIF14614 & Leiper* (A, BRI, CANB, GREE, K, L, MEL, MO, QRS); border track before Joahla Lookout, Lamington NP, 28°13′S, 153°11′E, *Forster PIF15114 & Leiper* (BRI); Green Mountain, O'Reilley's, Wishing Tree track, 28°14′S, 153°07′E, *Forster PIF15674 & Leiper* (BRI, MEL, NSW); Natural Arch NP 752, McPherson Ra., 28°14′S, 153°15′E, *Forster PIF16394* (BRI, QRS); Binna Burra, Mar 1941, *Gresty s.n.* (BRI AQ93465); [Lamington NP], Mt Wagawn, along walking track, 28°15′S, 153°13′E, *Guymer 2105* (BRI, CANB); midway between Araucarian Lookout and Mt Hobwee, Lamington NP, 28°15′S, 153°13′E, *Guymer 2111* (BRI, MEL); Upper Coomera R., *Johnson 139* (NSW); Upper Coomera R., May 1951, *Johnson s.n.* (NSW); Repeater Stn Rd, Springbrook, 28°14′S, 153°16′E, Jan 1993, *Leiper s.n.* (BRI AQ562124); SF 344 Mudgeeraba (Austinville SF), 28°S, 153°E, *McDonald et al. 2474* (BRI); vicinity of Picnic Rock, Lamington NP, Jan 1984, *Olsen s.n.* (BRI, CANB, DNA, MEL, MO); Picnic Rock, Lamington NP, 28°S, 153°E, Jul 1985, *Olsen s.n.* (BRI, CAN, GREE, MEL); Roberts Plateau, *Shirley 13* (NSW); Lamington NP, on Dave's Ck track, on Ck near junction with Ship's Stern track, 28°S, 153°E, *Smith 13899* (BRI); O'Reilly's, Lamington NP, *Smith & Webb 3620* (BRI, CANB); Austinville Rd, 7.5 km from turnoff to Springbrook Rd where road directly crosses Ck for first time, 28°11′S, 153°18′E, *Snow & Bradford 7463* (BRI, DNA, GREE, K, L, MO, NSW, US); Springbrook NP, adjacent to Blackfellows Ck, ca. 100 m upstream from trail leading below Canyon Lookout, 28°11′S, 153°17′E, *Snow & Bradford 7466* (BRI, CANB, GREE, K, MO, NSW); Morans Ck Paddock, Lamington Plateau, *Whaite & Whaite 3034* (NSW); Roberts Plateau, Lamington NP, Jan 1919, *White s.n.* (BRI, MEL,

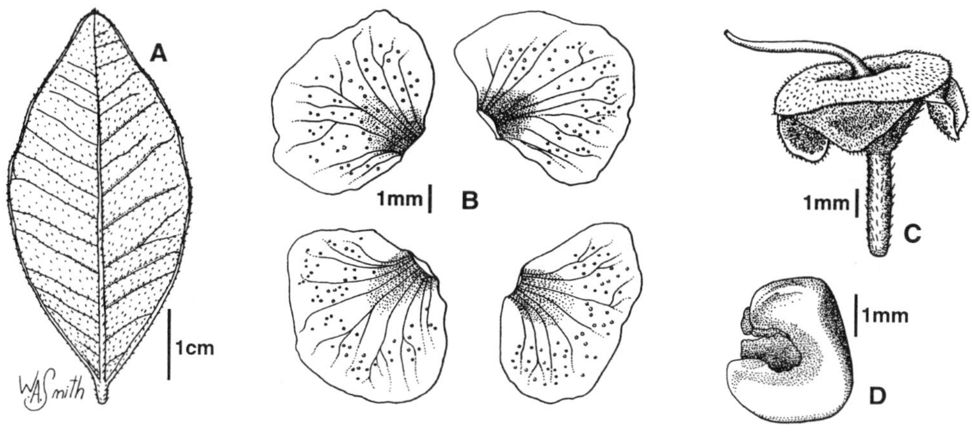

FIG. 8. *Lenwebbia prominens*. A. Leaf, abaxial view, showing prominent secondary veins. B. Petals. C. Hypanthium, calyx lobes, and style. D. Seed. (Based on: A, *Byrnes 4069*; B, C, *Leiper BRI 562124*.)

NSW); Roberts Plateau, Lamington NP, Mar 1920, *White s.n.* (BRI 93464).—NEW SOUTH WALES: Minyon Falls, N of Lismore, Sep 1929, *Cheel s.n.* (NSW 144687); Whian Whian SF, *Constable 22349* (NSW), *22358* (NSW, NY, US); Repentance Ck, Minyon Falls, N of Lismore, 28°37′S, 153°24′E, Jan 1985, *Guymer s.n.* (BRI, CANB, GREE, L, K, MO, NSW); Whian Whian, N of Lismore, 1945, *Hayes s.n.* (BRI AQ93455); MacPherson Ra., Jun 1957, *Hayes s.n.* (BRI AQ93454); Repentance Ck, Whian Whian S., SW of Mullumbimby, *Hayes et al. 2636* (NSW); Tweed Ra., Mebbin SF (7 mi SW of Tyalgum), Jun 1957, *Johnson & Constable s.n.* (BRI AQ93458, NSW 47037); along the Austinville Rd, near Lismore, 28°S, 153°E, *Leiper 5* (BRI); Whian Whian SF, 28°S, 153°E, 1953–1958, *Webb s.n.* (BRI AQ478287); Brindle Ck Track, Wiangarie SF, NE of Kyogle, 28°22′S, 153°03′E, Aug 1979, *Williams s.n.* (BRI, CANB); Whian Whian, 28°S, 153°E, Jun 1945, collector unknown (BRI AQ93561).

CULTIVATED. Cultivated [Bumdamba, Ipswich] from seed gathered at Springbrook, 27°S, 152°E, Nov 1980, *Bird s.n.* (BRI AQ344955); cultivated at Bundamba, 27°36′S, 152°49′E, Nov 1985, *Bird s.n.* (BRI, CANB, K, L, MO, NSW); Bundamba, Ipswich, property of L. Bird, 27°S, 152°E, Nov 1980, *Williams 80129* (BRI); Bundamba, cultivated (origin unknown), 27°36′S, 152°49′E, *Williams 87000* (BRI).

The oil glands on the abaxial leaf surface vary from not visible to the naked eye to prominent and abundant. The prominent and often raised secondary veins on the abaxial leaf surface consistently separate this species from the more common *L. lasioclada*.

Lenwebbia prominens is considered a useful species in landscaping (Wrigley & Fagg 1996).

The specific epithet is from the Latin *prominens* (prominent, standing or jutting out), in reference to the pronounced secondary veins of the lower leaf surface.

Gossia N. Snow & Guymer, gen. nov.—TYPE: *Gossia retusa* N. Snow & Guymer.

Arbores fruticesve erecti; flores (3–) 4–5 meri; parietes loculorum integri sine aperturis prope apicem, non ut in Myrto; filamenta pro antheris breviora quam in Myrto; ovula pro parte maxima 5–30; fructi maturi plerumque purpureonigri; embryones pro parte maxima circinati.

Erect multi-stemmed shrubs to large trees. Bark texture and color of trunk highly variable. Branchlets rounded, laterally compressed, or distinctly 4-angled. Leaves glossy to matte above, mostly matte below, decussate, opposite or rarely disjunct-opposite, more or less evenly distributed on branches, coriaceous. Stipules variable (sometimes even within a species), consisting of relatively short, thick, deep red, often laterally fused scales; of relatively stout ferrugineous setose hairs; or of basally swollen, relatively thick deep red hairs. Petioles channeled or smooth on abaxial surface. Leaf blades pinnately veined, the base cuneate, rounded, or cordate; oil glands sometimes present adaxially and abaxially. Inflorescence terminal or axillary, single, paired, or fascicled; flowers solitary, in racemes, or panicles, anthopodia and metaxphylls present or absent. Bracteoles subtending flowers usually 2 or infrequently 1 or 0, often falling early. Hypanthium tube cupulate to obconic, not extending beyond ovary apex. Sepals (3–) 4–5, fused proximally. Calyx lobes separate in bud, green or sometimes petaloid (greenish white), mostly persistent in fruit and held above body of fruit. Petals (3–) 4–5, concave, whitish or yellowish white. Stamens numerous, multiseriate, folded centerwards in bud; anthers mostly basifixed, versatile, dehiscing by longitudinal slits, tip of connective often with a single gland. Style straight, the stigma terete and relatively unswollen relative to diameter of style. Fruit a fleshy berry, yellow-orange or reddish to (usually) dark purple or nearly black at maturity. Ovary 1, locules 2; placenta capitate (often irregularly so), placentation axile; ovules (2–) 5–30 per placenta. Seed coat usually hard and bony, less frequently thickly membranous and somewhat leathery; adjacent seeds not fused. Embryo circinate or sometimes C-shaped, starchy or rarely oily on surface, with oil glands present or absent, endosperm mostly lacking or scantly present and gelatinous; hypocotyl thin to thick, the tip swollen or not, at same horizontal plane as embryo or rarely raised above plane. Seedling stems round, flattened, or winged. Chromosome number unknown.

Gossia is a genus of approximately 30 species and contains the majority of species formerly placed in *Austromyrtus* (Burret 1941; Stanley & Ross 1986; Henderson 1997; Hyland et al. 1999).

Gossia is most diverse in Australia and New Caledonia (Burret 1941; Snow 2000), although the predominantly Australian *G. floribunda* also occurs in New Guinea. Collectively, the genus is found in the Malesian, Northeast Australian, Neocaledonian (Guillaumin 1938), and Fijian (Guillaumin 1931) floristic regions of Takhtajan (1986). Within Australia, *Gossia* occurs in the Cape York, Atherton, Eastern Queensland, and McPherson-Macleay areas of endemism (Crisp et al. 1999). Most species grow in somewhat mesic habitats, such as moist to wet forests or along streambeds.

The generic name *Gossia* honors Wayne Goss (b. 1951), who approved several important biological and conservation initiatives for Queensland during his tenure as Premier of Queensland (1989–1996), including the construction of the new building in which the Queensland Herbarium is currently housed.

Key to the Species of Gossia in Australia

I. Key for Fertile Specimens

[Note: Lead 9 is most subject to misinterpretation.]

1. Flowers predominantly 5-merous.
    2. Branchlets densely hairy.
        3. Peduncles 2–7 mm long; midrib of upper leaf surface raised slightly; oil glands absent on upper leaf surface; oil glands on lower surface sparse. 6. *G. inophloia*.
        3. Peduncles 2 mm or less long; midrib of upper leaf surface flush; oil glands visible on upper surface; oil glands on lower surface common. 7. *G. pubiflora*.
    2. Branchlets becoming glabrous to slightly sericeous.
        4. Bark of trunk generally very smooth.

5. Oil glands on upper leaf surface usually abundant; petiole usually channeled, often deeply so; calyx lobes whitish or greenish white. 8. *G. bidwillii.*
5. Oil glands on upper leaf surface sparse to common; petiole smooth above to slightly channeled (rarely deeply so); calyx lobes greenish. 9. *G. acmenoides.*
4. Bark of trunk usually somewhat rough.
6. Adaxial midrib of leaf raised.
7. Branchlets winged. 10. *G. gonoclada.*
7. Branchlets wingless. 11. *G. hillii.*
6. Adaxial midrib of leaf flush.
8. Leaf tip (see Methods) pointed; stipules ochreate, short and generally faint, sheathing the branchlet of younger leaves; midveins of abaxial leaf reddish brown; oil glands on abaxial leaf visible only with magnification; hairs on petals ferrugineous; anther sacs sometimes with minute oil glands. 12. *G. lewisensis.*
8. Leaf tip rounded; ochreate stipules absent; midveins of abaxial leaf greenish; oil glands on abaxial leaf visible without magnification; hairs on petals whitish; anther sacs without minute oil glands. 13. *G. punctata.*
1. Flowers predominantly 4-merous.
9. Abaxial leaf surface glossy.
10. Branchlets winged or quadrangular.
11. Leaf apex retuse; base cuneate to rounded. 24. *G. retusa.*
11. Leaf apex acute to acuminate; base cuneate. 20. *G. floribunda.*
10. Branchlets flattened or terete.
12. Stipules absent (check closely); bracteoles absent on most flowers; petioles (dried material) black or nearly so. 15. *G. grayi.*
12. Stipules present; bracteoles present on most flowers (except mostly absent on *G. bamagensis*); petioles (dried material) light or dark but never black.
13. Oil glands on adaxial leaf surface visible without magnification.
14. Branchlets puberulous; leaves often silvery-green above; fruit orangish yellow, turning red at maturity; bark rough, fissured or flaky. 22. *G. fragrantissima.*
14. Branchlets glabrous; leaves usually dark green above; fruit red to dark blue to black (at maturity); bark smooth. 19. *G. lucida.*
13. Oil glands on adaxial leaf surface visible only with magnification.
15. Oil glands on abaxial leaf surface often visible without magnification; bracteoles below flowers absent (rarely 1 or 2) or as undifferentiated protrusions; oil glands on petals fewer than 30. 17. *G. bamagensis.*
15. Oil glands on abaxial leaf surface not or only barely visible (even at 10×); bracteoles below flowers usually two; oil glands on petals 100–175. 18. *G. shepherdii.*
9. Adaxial leaf surface usually matte.
16. Longer leaf blades mostly 10–20 cm long; innermost intramarginal veins on abaxial leaf surface 4–14 mm from margin. 14. *G. dallachiana.*
16. Longer leaf blades usually less than 10 cm long; innermost intramarginal veins on abaxial leaf surface 4 mm or less from margin.
17. Leaf tip rounded, sometimes retuse or mucronate, rarely falcate. 23. *G. sankowskiorum.*
17. Leaf tip obtuse to acuminate.
18. Branchlets quadrangular (check closely).
19. Petioles 1.5–4.5 mm long, distinctly channeled; leaf blades 1.5–7.5 cm long, 1.0–4.0 cm wide, base cuneate, apex mostly acuminate; inflorescence a monad borne on pubescent short shoots; filaments glabrous. 20. *G. floribunda.*
19. Petioles 1.0–1.5 mm long, unchanelled or only slightly so; leaf blades 1.7–3.2 cm long, 0.9–1.6 cm wide, base rounded to slightly cordate, apex obtuse to acute; inflorescence a monad or consisting of 2 flowers on glabrous peduncles; filaments sparsely sericeous. 25. *G. byrnesii.*
18. Branchlets terete or somewhat flattened.
20. Stipules absent (check closely); bracteoles absent on most flowers; petioles (dried material) black or nearly so. 15. *G. grayi.*
20. Stipules present; bracteoles present on most flowers; petioles light to dark but never black.

21. Oil glands of adaxial leaf surface visible without magnification. 18. *G. lucida.*
21. Oil glands of adaxial leaf surface not visible or requiring magnification.
 22. Leaf apex obtuse (or sometimes slightly acuminate). 16. *G. macilwraithensis.*
 22. Leaf apex acuminate with a rounded or acute tip.
  23. Trunk not fluted or buttressed; bark rough, scaly or somewhat fissured; petioles at least slightly channeled; oil glands abundant on adaxial and abaxial leaf surfaces (use magnification); hypanthium 1.0–2.5 mm long; calyx lobes 0.7–2.5 mm long, upper surface sparsely sericeous near margins; petals 2.5–5.5 mm long. 21. *G. myrsinocarpa.*
  23. Trunk sometimes fluted or buttressed; bark smooth, thin; petioles usually unchannelled; oil glands on adaxial leaf surface sparse (use magnification) but common on abaxial surface; hypanthium 1.7–4.5 mm long; calyx lobes 1.8–3.0 mm long, upper surface sericeous to densely sericeous; petals 4.0–9.5 mm long. 18. *G. shepherdii.*

II. Key for Sterile Specimens

NOTE: Sterile specimens of *Gossia* can resemble some species of *Decaspermum*, *Uromyrtus*, *Archirhodomyrtus*, *Acmena* DC., *Eugenia* L., and *Syzygium* Gaertn. The following key works for most sterile material, particularly if the sample has young branchlets and bark samples or descriptions from the main trunk. When coupled with known geographic ranges, the majority of material should be identifiable. Some species are keyed out more than once. For species occurring in the tropical rainforests of northeastern Queensland see also Hyland et al. (1999) under *Austromyrtus*.

1. Youngest leaves with short, faint, ochreate (sheathing) stipules. 12. *G. lewisensis.*
1. Youngest leaves without ochreate stipules.
 2. Branchlets quadrangular or with four narrow wings (two on either side of somewhat flattened internode).
  3. Midrib of adaxial leaf surface flush. 25. *G. byrnesii.*
  3. Midrib of adaxial leaf surface prominently raised near base.
   4. Stipular hairs bulbous at base and some arising lateral to base of petiole. 20. *G. floribunda.*
   4. Stipular hairs neither bulbous at base nor arising lateral to base of petiole.
    5. Plants usually less than 5 m tall; stipules consisting of several dark red, short, laterally fused, scalelike structures, which often produce a resinous exudate in leaf axils; leaves mostly 1.5–3.0 cm long, the margins slightly but tightly revolute; leaf apex obtuse. 24. *G. retusa.*
    5. Plants often more than 5 m tall; stipules consisting of setose hairs, sometimes swollen at base, not producing a resinous exudate in leaf axils; leaves mostly 3.0–6.0 cm long, the margins mostly flat; leaf apex acute. 10. *G. gonoclada.*
 2. Branchlets terete, without wings.
  6. Branchlets with moderate to dense covering of hairs.
   7. Midvein of upper leaf surface slightly raised throughout. 6. *G. inophloia.*
   7. Midvein of upper leaf surface flush.
    8. Some hairs on branchlets more than 0.5 mm long. 7. *G. pubiflora.*
    8. All hairs on branchlets less than 0.5 mm long. 22. *G. fragrantissima.*
  6. Branchlets mostly glabrous to sparsely sericeous.
   9. Midribs of upper leaf surface raised towards base.
    10. Stipules absent (check closely); petioles of dried material black or nearly so. 15. *G. grayi.*
    10. Stipules present; petioles of dried material various, but usually not blackish.
     11. Innermost intramarginal vein on lower leaf surface 4–14 mm from margin; larger leaf blades 85–200 mm long. 14. *G. dallachiana.*
     11. Innermost intramarginal vein on lower leaf surface < 4 mm from margin; larger leaf blades mostly < 100 mm long.
      12. Upper leaf surface somewhat glossy or matte; bark smooth. 9. *G. acmenoides.*
      12. Upper leaf surface very glossy; bark rough. 11. *G. hillii.*

9. Midribs of upper leaf surface flush throughout.
 13. Youngest leaves with short (and faint) ochreate (=sheathing) stipules; midveins of lower leaves mostly reddish brown. 17. *G. lewisensis.*
 13. Youngest leaves without ochreate stipules; midveins of lower leaves not reddish brown.
  14. Oil glands of upper leaf surface usually not visible to the naked eye.
   15. Leaf base rounded or slightly cordate *and* petioles less than 2 mm long.
    16. Nodes of branchlets swollen; oil glands on lower leaf surface sparse to common; leaf apex obtuse (or sometimes slightly acuminate); stipular hairs setose and often extending lateral to base of petiole (often not visible). 16. *G. macilwraithensis.*
    16. Nodes of branchlets not swollen; oil glands on lower leaf surface dense; leaf apex obtuse to acute with a rounded tip; stipular hairs basally swollen. 9. *G. byrnesii.*
   15. Leaf bases cuneate *and/or* petioles 2 mm or longer.
    17. Bark of trunk rough.
     18. Upper leaf surface glossy, with oil glands sparse (use 30× magnification), lateral and intramarginal veins absent on lower leaf surface; petiole 1–3 mm. 13. *G. punctata.*
     18. Upper leaf surface matte, with oil glands abundant; lateral and intramarginal veins on lower leaf surface visible; petioles 3–7 mm. 21. *G. myrsinocarpa.*
    17. Bark of trunk smooth.
     19. Plants from northern New South Wales to central Queensland. 6. *G. acmenoides.*
     19. Plants from northeastern Queensland.
      20. Bark of trunk thin, "python"-colored with shaded patches of browns, grays, reds, or greens. 18. *G. shepherdii.*
      20. Bark of trunk sometimes conspicuously layered, reddish, orangish, or somewhat greenish. 17. *G. bamagensis.*
  14. Oil glands of upper leaf surface visible to the naked eye.
   21. Leaf apex retuse. 23. *G. sankowskiorum.*
   21. Leaf apex acute to acuminate.
    22. Petiole channeled (sometimes deeply); leaf base rounded or occasionally cuneate; oil glands on adaxial leaf surface abundant and often dark. 8. *G. bidwillii.*
    22. Petiole smooth (or sometimes deeply channeled); leaf base cuneate or less often rounded; oil glands on adaxial leaf surface often abundant but not typically dark.
     23. Bark of trunk brown or gray and sometimes with patches of orange or green; petioles smooth on upper surface. 19. *G. lucida.*
     23. Bark of trunk reddish, orangish, or sometimes somewhat greenish; petioles smooth or sometimes deeply channeled. 17. *G. bamagensis.*

6. Gossia inophloia (J. F. Bailey & C. T. White) N. Snow & Guymer, comb. nov. *Myrtus inophloia* J. F. Bailey & C. T. White, Queensl. Dept. Agric. Div. Pl. Industr. Bull. 19: 8. 1917. *Austromyrtus inophloia* (J. F. Bailey & C. T. White) Burret, Notizbl. Bot. Gart. Berlin-Dahlem 15: 502. 1941.—TYPE: AUSTRALIA. Queensland: Blackall Range, Nov 1916, *C. T. White s.n.* (lectotype, here designated: BRI!).

Myrtus decaspermoides Domin, Biblioth. Bot. 89: 1029. 1928. *Austromyrtus decaspermoides* (Domin) Burret, Notizbl. Bot. Gart. Berlin-Dahlem 15: 502. 1941.—TYPE: AUSTRALIA. Queensland: Moreton District, Eudlo, 9 Nov 1891, *F. M. Bailey s.n.* (holotype: PR!, photo: BRI!; isotype: BRI!).

Shrubs to treelike shrubs, 2.5–5 m tall. Bark of trunk smooth to stringy, brown to gray. Branchlets rounded to laterally compressed, not winged, brown to reddish brown, smooth but often becoming slightly stringy, densely sericeous to villous, oil glands absent or obscured by hairs. Leaves concolorous to slightly discolorous, slightly glossy above, matte below. Stipules type A (see Morphology). Petioles 1–2.5 mm long, brown to reddish, not channeled, eglandular. Leaf blades 0.7–3.5 (–4.7) cm long, (0.4–) 1.0–2.3 cm wide, elliptic to narrowly ovate; base cuneate to rounded; apex mostly acute or sometimes acuminate, the tip pointed to often mucronate; margins flat or somewhat revolute; adaxial surface sericeous but soon becoming glabrous, oil glands few or not visible to the naked eye; midvein raised slightly throughout; abaxial surface appressed-sericeous with ferrugineous hairs, oil glands few, visible only with magnification; secondary veins absent to indistinct; intramarginal veins usually absent or rarely faint; the inner intramarginal vein 1–1.5 mm from edge. Inflorescence terminal or axillary, solitary; monads (often clustered on short axillary branches) or in racemes; anthopodia and metaxyphylls absent. Peduncles 2–7 mm long, rigid, sparsely villous to villous. Bracteoles 2, 1.5–4 mm long, 0.4–0.5 mm wide, linear, irregularly flexuous to rigid, scalelike without midrib, with tips exceeding base of calyx lobes, sericeous to villous, caducous in fruit. Hypanthium 1.3–2 mm long, cupulate, densely sericeous to densely villous, oil glands few and obscured by hairs. Calyx lobes 5, 1.5–2.5 mm long, broadly triangular to broadly ovate, apex obtuse or acute, greenish; adaxial surface sericeous to villous above (particularly near apex); abaxial surface sparsely villous. Petals 5, 2–3 mm long, 1.5–2 mm wide, widely elliptic to circular, whitish; adaxial surface glabrous to sericeous near apex; abaxial surface sparsely villous to shortly villous, oil glands few to common. Stamens up to 75, included, filaments 2–3 mm long; staminal disk glabrescent; anthers ca. 0.5 mm long, subglobose, the tip of connective with a single (small) gland. Style 3–4 mm long, glabrous to sparsely sericeous near base; ovary apex slightly hairy; ovules 5–6 per placenta, placed irregularly. Berry 5–10 mm long, 4–9 mm in diameter, subcylindrical, base rounded to tapered, sparsely sericeous (often shortly so), dark blue-black. Seeds in mature fruit 1; subglobose, somewhat flattened, smooth, light brown, micropylar operculum faint or absent, seed coat hard, bony. Embryo circinate, starchy, oil glands absent, endosperm absent; hypocotyl relatively thick, tip near radicle swollen, tip at same plane as embryo; cotyledons circinate, shorter than hypocotyl, pointing forward.

Phenology. Flowering in November and December; fruiting March and April.

Distribution (Fig. 9). Southeastern Queensland in the northern part of the McPherson-Macleay area of endemism (Crisp et al. 1999); occasional to locally common in rainforests and mixed notophyll forests on ridges, slopes, or along streams; 120–720 m.

Conservation status. The species is considered Vulnerable based on criteria C2a (Species Survival Commission 1994; see also Forster et al. 1991.)

Vernacular name. Thread-barked myrtle.

ADDITIONAL SPECIMENS EXAMINED. **Australia.** QUEENSLAND. WIDE BAY DISTRICT: Panorama Drive via Duke Rd, 8 km SE of Cooroy, 26°27′S, 152°59′E, *Bean 5935* (BRI); Delicia Rd, 5 km W of Mapleton, 26°37′S, 152°49′E, *Bean 10178* (BRI); Petrie Ck, Dulong Rd, SE of Mapleton, 26°40′S, 152°56′E, *Guymer 2130* (BRI); Dikusha, Obi Obi Gorge, W of Montville Rd, 4.3 km ENE of Maleny, 26°44′S, 152°53′E, *Grimshaw PG809* (BRI); Mary Cairncross Park, ca. 4 km ESE of Maleny, 26°45′S, 152°53′E, *Henderson H3043 & Sharpe* (BRI); Kin Kin, *Shirley s.n.* (BRI 93449); Gully E of Sunrise Rd, Noosa Shire, S of SF 959, 26°25′S, 153°00′E, *Thomas 46* (BRI); Blackall Ra., Nov 1916, *White s.n.* (MEL 726888).—MORETON DISTRICT: Eudlo, Nov 1891, *Bailey & Simmonds s.n.* (MEL 726898); Mt Glorious, Jan 1945, *Clemens s.n.* (BRI AQ93452, US); Mt Glorious, near Maiala NP, 27°19′S, 153°45′E, *Forster PIF15870 & Leiper* (BRI); Mt Glorious, adjacent to Camp Constable, 27°20′S, 152°46′E, *Guymer 2119* (BRI); Mt Glorious, 27°S, 152°E, *Jessup & Byrnes 4062* (BRI); Mt Glorious, 27°S, 152°E, Nov 1972, *Lebler & Durrington s.n.* (BRI 14695, CANB, MEL 238757); North Maroochy R., 26°S, 153°E, 1984, *McCabe s.n.* (BRI AQ394774); property of BA & SP Phillips, Lot 37 Alex Rd, Mt Glorious, 27°19′S, 152°44′E, Dec 1992, *Phillips s.n.* (BRI AQ569021); Mt Glorious, 27°19′S, 152°45′E, *Snow et al.*

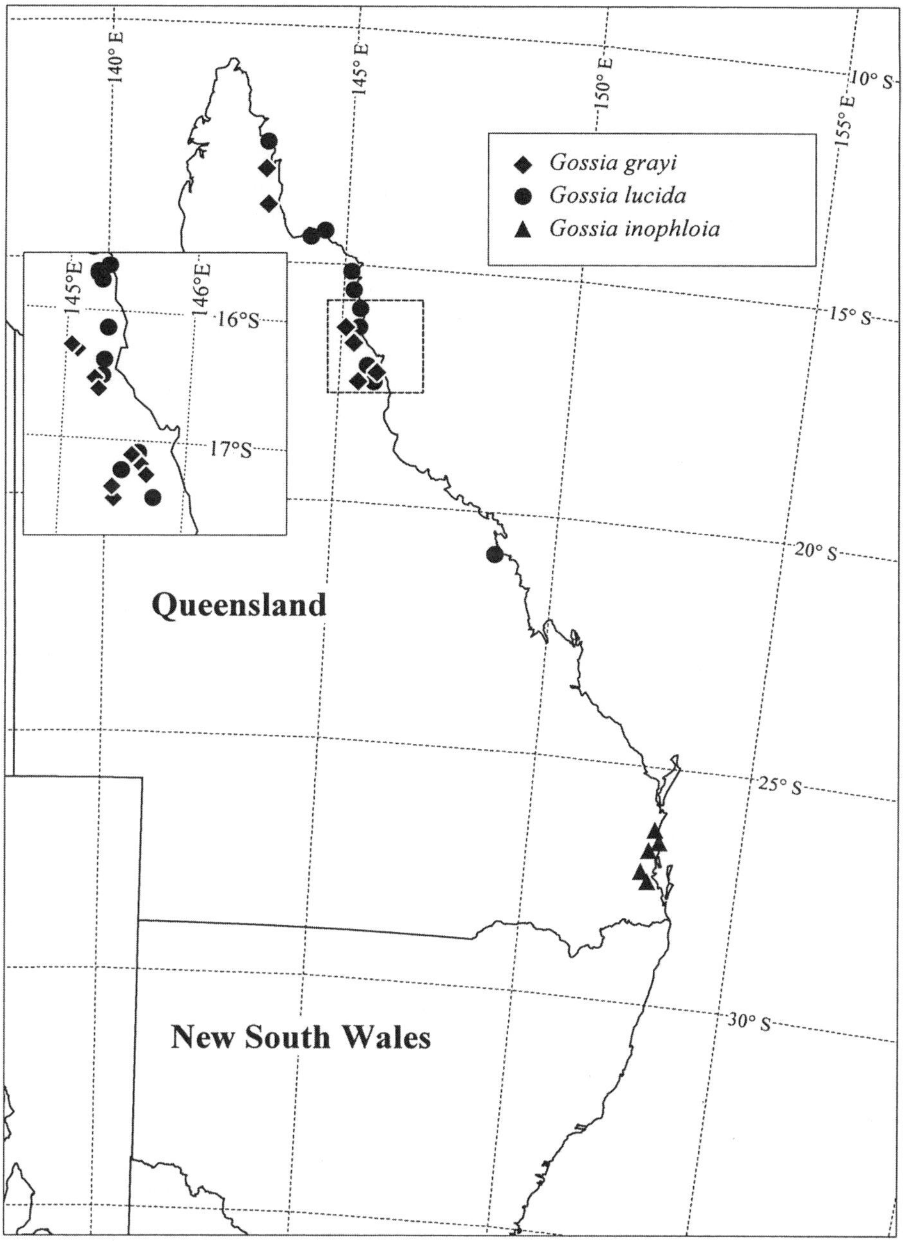

FIG. 9. Distribution of *Gossia grayi*, *G. inophloia*, and *G. lucida*.

7452 (BRI); along walking track, Mary Cairncross Park, Maleny, 26°46′S, 152°52′E, *Thomas RFR35 & Gale* (BRI); Macintyre's property, Rumwodt Rd, Dulong, W of Nambour, 26°39′S, 152°53′E, Jan 1993, *Thomas s.n. & Gayle* (BRI AQ636359); Mt Glorious, *White 11071* (BRI); Mt Glorious, *White 12075* (BRI); Peachester, May 1918, *White s.n.* (BRI 93447).

The ferrugineous hairs on the branchlets, raised midvein of the upper leaf, and bracteole apices that generally extend slightly beyond the base of the sepal lobes collectively distinguish *G. inophloia* from the similar *G. pubiflora*. The latter species is more densely sericeous on the calyx lobes and bracteoles.

The specific epithet derives from the Greek *ino* (thread) and *phloios* (bark) and refers to the often stringy bark of the species.

7. Gossia pubiflora (C. T. White) N. Snow & Guymer, comb. nov. *Myrtus pubiflora* C. T. White, Proc. Roy. Soc. Queensland 50: 77. 1939. *Austromyrtus pubiflora* (C. T. White) L. S. Smith, Proc. Roy. Soc. Queensland 67: 35. 1956.—TYPE: AUSTRALIA. Queensland: North Kennedy District, Strathdickie, Proserpine District, *N. Michael 1476* (holotype: BRI!).

Compact to multi-stemmed shrubs or trees, 1–6 m tall. Bark of trunk rough, flaky, scaly or tessellated, light brown to gray. Branchlets rounded, not winged, light to dark brown, smooth, puberulous (some hairs more than 0.5 mm long), oil glands few, indistinct. Leaves concolorous or slightly discolorous, matte above and below. Stipules absent, or type A and obscure (see Morphology). Petioles (1–) 2.5–4 mm long, green to dark brown, often slightly channeled, eglandular. Leaf blades 1.8–5.5 cm long, 0.8–3.3 cm wide, elliptic to ovate; base cuneate to rounded; apex acute; the tip pointed and often mucronate; margins flat; adaxial surface glabrous or sparsely tomentose along lower edges, oil glands common, faint or visible with magnification, midvein flush; abaxial surface glabrous or sometimes sparsely sericeous near base of midvein, oil glands common, visible with magnification; secondary and intramarginal veins indistinct to prominent; the inner intramarginal vein 0.5–2 (–3) mm from edge. Inflorescence terminal or axillary, solitary or paired, a monad or occasionally a triad; anthopodia and metaxyphylls absent. Peduncles to 2 mm long, rigid, nearly sessile, villous. Bracteoles 2, 2–2.5 mm long, 0.5–0.6 mm wide, linear to narrowly triangular, ascending to rigid, scalelike without midrib, with tips shorter than base of calyx lobes, villous to densely villous, persistent or caducous in fruit. Hypanthium 4.5–5.5 mm long, cupulate, villous to densely villous, oil glands few. Calyx lobes (4–) 5, 5–8 mm long, triangular, apex acuminate, greenish; both surfaces sericeous to villous. Petals 5, 4–5 mm long, ca. 4 mm wide, obovate to oblate or circular, yellowish white; adaxial surface glabrous; abaxial surface shortly villous, oil glands few. Stamens up to 150, included; filaments 1.8–2.3 mm long; staminal disk glabrous; anthers 0.6–0.8 mm long, subglobose to globose, the tip of connective with a single gland. Style 2.5–3 mm long, glabrous; ovary apex glabrous; ovules 15–18 per placenta, placed irregularly. Berry 9–17 mm long, 11–19 mm wide, globose to subglobose; base rounded to slightly tapered, glabrescent, red to nearly black. Seeds in mature fruit 1; rounded, smooth, medium to dark brown color, micropylar operculum present, seed coat hard, bony. Embryo circinate, oil glands absent, starchy; endosperm absent or present in small amounts; hypocotyl relatively thick, tip near radicle swollen, tip at same plane as embryo; cotyledons circinate or mostly straight, shorter than hypocotyl, pointing forward.

Phenology. Flowering in August and September; fruiting throughout the year.

Distribution (Fig. 16). East central Queensland in the Eastern Queensland area of endemism (Crisp et al. 1999); occasional to common in rainforests or notophyll vineforests, often on slopes; 50–200 m.

Conservation status. The species is considered Vulnerable (criteria C2a, Species Survival Commission 1994).

ADDITIONAL SPECIMENS EXAMINED. **Australia**. QUEENSLAND. NORTH KENNEDY DISTRICT: Conway NP, rear of barracks, 20°17′S, 148°45′E, *Champion 1005 & Pollock* (BISH, BRI); Shute Harbour, Dec 1979, *Cribb s.n.* (BRI AQ331198); Vine Ck, W slopes of Mt Dryander, 20°14′S, 148°33′E, *Guymer 1721* (BRI); Dryander Ck, 4 km ESE of Mt Dryander, 20°16′S, 148°36′E, *Guymer 2004* (BRI, CANB); 4.6 km along Brandy Ck Rd, S of Shute Harbour-Proserpine Rd, 20°21′S, 148°41′E, *Guymer 2078* (BISH, BRI, CANB, MO); SFR 299, Conway, 20°20′S, 148°45′E, *Hyland 4070* (QRS); ca. 25 km NE of Proserpine, SE of Airlie, Campbell Ck Valley, 20°17′S, 148°44′E, *Kanis 2133* (CANB, MO, QRS); Palm Bay to sandy Bay Track, Long Is., 20°21′S, 148°50′E, *McDonald 5621 & Bean* (BRI); Conway NP, vicinity of camping area, 20°17′S, 148°45′E, *McDonald et al. 4533* (BRI); Strathdickie North, *McPherson 82* (QRS); Conway SF, Sep 1994, *Perry s.n.* (BRI AQ632036); Dryander Ck, ca. 2 km N of Gregory and ca. 20 km N of Proserpine, 20°24′S, 148°35′E, *Sharpe 4186* (BRI); Brandy Ck Rd, 4.5 km E of Shute Harbour Rd and 13 km NE of Proserpine, 20°20′S, 148°42′E, *Sharpe 4040 & Perry* (BRI, CANB).

The attenuate calyx lobes, densely hairy branchlets and flowers, and relatively large fruits make this species unmistakable in *Gossia*. Mature fruits are said to have a menthol smell when crushed. The majority of collections seen are sterile.

The specific epithet derives from the Latin, meaning softly hairy flowers.

8. Gossia bidwillii (Bentham) N. Snow & Guymer, comb. nov. *Myrtus bidwillii* Bentham, Fl. austral. 3: 275. 1867. *Austromyrtus bidwillii* (Bentham) Burret, Notizbl. Bot. Gart. Berlin-Dahlem 15: 501. 1941.—TYPE: AUSTRALIA. Queensland: Wide Bay District, Wide Bay, Northeast Australia, *J. C. Bidwill s.n.* (holotype: K!; photo: BRI!).

Myrtus racemulosa Bentham, Fl. austral. 3: 276. 1867. *Austromyrtus racemulosa* Burret, Notizbl. Bot. Gart. Berlin-Dahlem 15: 501. 1941.—TYPE: AUSTRALIA. Queensland: North Kennedy District, Edgecombe Bay, *J. Dallachy s.n.* (lectotype, here designated, MEL!; isolectotype: K!, photo: BRI!).

Myrtus racemulosa var. *conferta* Bentham, Fl. austral. 3: 276. 1867.—TYPE: AUSTRALIA. Queensland: North Kennedy District, Port Denison, *E. F. A. Fitzalan s.n.* (holotype: K!, photo: BRI!; isotypes: K! MEL-92365!, photo of K isotype: BRI!).

Trees or sometimes shrubs, 2–15 m tall. Bark of trunk usually smooth and often with thin flakes, often "python"-colored with patches of white, brown, pink, or green. Branchlets rounded, not winged, whitish to brown, smooth, glabrous, oil glands few to common, evidently absent or faint to prominent. Leaves concolorous to discolorous, matte to glossy above, matte to glossy below. Stipules mostly type C or sometimes type B (see Morphology). Petioles 2.5–5 (–9) mm long, light green to dark brown, channeled (often deeply so), eglandular. Leaf blades 3.0–9.0 (–10.5) cm long, 1.5–4.5 (–6.0) cm wide, elliptic to ovate; base rounded or occasionally cuneate; apex acute to acuminate, the tip pointed or sometimes rounded, margins flat; adaxial surface glabrous, oil glands abundant and often dark, not visible to the naked eye to prominent, midvein flush or rarely slightly raised; abaxial surface glabrous, oil glands common to abundant, not visible to the naked eye but prominent when viewed with magnification; secondary veins not visible to the naked eye to prominent, the intramarginal veins invisible to indistinct, inner intramarginal vein 1–4 (–7) mm from edge. Inflorescences terminal or axillary, often paired or fascicled; occasionally a single flower or more frequently a raceme or panicle; anthopodia present or absent, metaxyphylls absent or present (and less than 1 mm long). Peduncle 1.5–17 mm long, rigid or slightly flexuous, glabrous. Bracteoles 2, 0.5–1.5 mm long, 0.2–0.5 mm wide, linear to narrowly triangular, stiff, scalelike without midrib, the tips shorter than base of calyx lobes, glabrous to sparsely sericeous, mostly caducous in fruit. Hypanthium 0.7–2.5 mm long, obconic to cupulate, glabrous to sparsely sericeous, oil glands few to abundant. Sepal lobes 5 (rarely 4), 0.5–1.8 mm long, broadly ovate to rounded, usually whitish or greenish white, apex obtuse; adaxial surface glabrous to sericeous (especially on margins and near apex); abaxial surface glabrous. Petals 2–5.5 mm long, 2.5–5 mm wide, widely elliptic to circular, white to yellowish white; adaxial surface glabrous to sparsely sericeous near apex; abaxial surface glabrous, oil glands common. Stamens (55–) 75–120, included, filaments 1.5–5 mm long; staminal disk glabrous to slightly glabrescent; anthers 0.5–0.9 mm long, subglobose to globose, the tip of connective eglandular or with a single gland. Styles 1.5–5.5 mm long, glabrous to sparsely sericeous below; ovary apex glabrous to hairy; ovules 8–25 per placenta, placed irregularly. Berry (4–) 6–10 mm long, 5–13.5 mm in diameter, subglobose to globose, base rounded, glabrous, dark purple to black. Seeds in mature fruit 2–3; rounded to reniform, somewhat flattened, smooth to slightly wrinkled, light brown, micropylar operculum present or only weakly formed, seed coat mostly hard but sometimes leathery or very thin. Embryo circinate, starchy, oil glands present (sometimes as small papillae), endosperm absent or present; hypocotyl relatively thick, tip near radicle swollen, tip at same plane as embryo; cotyledons mostly straight, shorter than and surrounded by circinate hypocotyl, pointing forward. Fig. 10.

Phenology. Flowering from August through January, peaking October and November; fruiting October through March.

Distribution (Fig. 11). Northern New South Wales throughout eastern Queensland, mostly east of the Great Dividing Range, including the McPherson-Macleay, Eastern Queensland, Atherton, and Cape York areas of endemism; occasional to very common on ridgetops and slopes in Araucarian microphyll vineforests, complex notophyll forests, semi-evergreen vineforest thickets, dry scrub, and gallery rainforests; on a variety of substrates, including serpentinites (*Batianoff 9812145*, BRI); 70–1100 m.

Vernacular name. Python tree, smooth-barked ironwood, scrub ironwood, lignum, lignum-vitae.

ADDITIONAL SPECIMENS EXAMINED. **Australia.** QUEENSLAND. COOK DISTRICT: Yarramulla Stn, left side of Gulf Developmental Rd, 100 km SSW of Mt Garnet, 18°10′S, 144°40′E, *Batianoff 900144 & Smith* (BISH, BRI); Atherton, 17°15′S, 145°30′E, Oct 1972, *Bravery s.n.* (QRS 3090); Rooty Ck, near Monamona Mission Stn, 16°43′S, 145°30′E, *Flecker 6329* (BRI, QRS); Mt Windsor Tableland, SFR144, 9 km past Spencer Ck Crossing, 16°18′S, 145°05′E, *Forster PIF13701 et al.* (BRI, GREE, MEL, QRS); SF 191 Wongabel, near Barron R., 17°19′S, 145°31′E, *Forster PIF 14447* (BRI); SFR 607 Shoteel LA Mickies Pocket, 16°55′S, 145°36′E, *Gray 2319* (BRI, QRS); SFR 607 Shoteel LA Mickies Pocket, 16°56′S, 145°37′E, *Gray 2781* (BRI); 62 km S of Mt Garnet along Kennedy Hwy, 40 Mile Scrub, 18°06′S, 144°50′E, *Halford Q776* (BRI, MEL); TR176, Monkhouse, 15°44′S, 145°14′E, *Hyland 12096* (BRI); SFR185, Danbulla, 17°09′S, 145°38′E, *Moriarty 1155* (BRI, CANB); Robson LA, E. P. 2, 17°36′S, 145°38′E, *Risley 65* (BRI, CANB, QRS); Danbulla, Stony Ck LA, 17°09S, 145°35′E, *Smith 10123* (BRI); Downfall Ck near Tinaroo Dam (Danbulla SF), 17°S, 145°E, *Tucker 52* (BRI); Danbulla, 17°09′S, 145°37′E, Oct 1969, *Volck s.n.* (BRI AQ93371).—NORTH KENNEDY DISTRICT: Hook Is., Whitsunday region, 20°S, 148°E, *Batianoff 3608* (BRI); Double Cone Is., Whitsunday region, 20°06′S, 148°43′E, *Batianoff 3672* (BRI, CANB); Daydream Is., Whitsunday region, 20°15′S, 148°53′E, *Batianoff 900337* (BISH, BRI); Cape Cleveland, Bowling Green Bay NP, S of Townsville, 19°16′S, 147°02′E, *Bean 5308*

(BRI); Bowen, 20°01′S, 148°15′E, *Blake 18522* (BRI, CANB, MO, NSW); Forty Mile Scrub NP, 3.8 km N of the Mt Surprise road junction on the Kennedy Hwy, 18°06′S, 144°49′E, *Clarkson 6901 & McDonald* (BRI); The Bluff, E of Mingela, ca. 70 km S of Townsville, 19°53′S, 146°44′E, *Cumming 9342* (BRI); Port Denison, Aug 1863, *Fitzalan s.n.* (MEL 92366, 92370–92372); Forty Mile Scrub, S of Mt Garnet, 18°07′S, 144°49′E, *Forster PIF9640* (A, B, BRI, GREE, L, K, MEL, NY); Forty Mile Scrub, W of Mt Garnet, *Fosberg 61619* (BRI); Undara Lava Tunnel, SE of Mt Surprise, 18°13′S, 144°36′E, *Godwin C3256* (BRI, QRS); Castle Hill, near Townsville, 1886, *Gulliver s.n.* (MEL 726885); 6.8 km SE Bowen, 20°S, 148°E, *Guymer 1125* (BRI); 40 Mile Scrub, 18°05′S, 144°50′E, *Hyland 11637* (BRI, QRS); Magnetic Is., West Point, *Jackes 9001* (MEL); Cleveland Bay, 1876, *Johnson s.n.* (MEL 221899 and 221900); slopes of Flame Tree Hill, ca. 3 km ESE of Airlie Beach, 20°17′S, 148°44′E, *McDonald 5846* (BRI); Main Ck draining into Double Bay, Dryander NP, ca. 6 km SSE of Earlando, 20°12′S, 148°34′E, *McDonald et al. 5792* (BRI, CANB, NSW); Strathdickie North, near Proserpine, *McPherson 87* (BRI, QRS); Edgecomb Bay, *Michael s.n.* (BRI AQ93346); Hwy 1, 70 mi W of Ravenshoe, *Riggins 1288* (RSA); Hayward Ck, Conway NP, Shute Harbour ca. 35 km NE of Proserpine, 20°18′S, 148°45′E, *Sharpe 4144* (BRI, CANB); Rocky Hill and base, 20°S, 147°E, *Smith 4516* (BRI, CANB), *4516A* (BRI); Forty Mile Scrub, 18°07′S, 144°49′E, *Stocker 1785* (BRI); SF R461, Five Mile Ck, ca. 6.5 km S of Cardwell, 1.5 km W of Bruce Hwy, *Thornsborne et al. 491* (BRI, CANB); SF 461, Five Mile Ck, ca. 6.5 km S of Cardwell, 1.5 km W of Bruce Hwy, *Thornsborne et al. 478* (BRI, CANB); Headwaters of Dryander Ck, Mt Dryander, ca. 20°15′S, 148°35′E, *Webb & Tracey 10057* (BRI, CANB, MEL, NSW); 3 mi from 40 Mile Scrub S of Mt Garnet, *Wyatt 8* (BRI).—SOUTH KENNEDY DISTRICT: Shaw Is., 20°29′S, 149°09′E, *Batianoff 3196* (BRI, CANB, MO); Broad Sound, Sep 1802, *Brown s.n.* (NSW); Mt Beatrice NP, 3 km SW of Wagoora, 20°52′S, 148°40′E, *Forster PIF14294* (BRI, MEL).—LEICHHARDT DISTRICT: Bauhinia Ck, 12.3 km E of Harold Hinchcliffe Bridge, Rockhampton and Duaringa (24.8 km E of rest area on E side of Duaringa), 23°45′18″S, 149°52′18′E, *Brown 92/12 & Doust* (ASU, NSW); Glenmoral Gap, 13 km WSW of Theodore, 24°58′S, 149°57′E, *Forster*

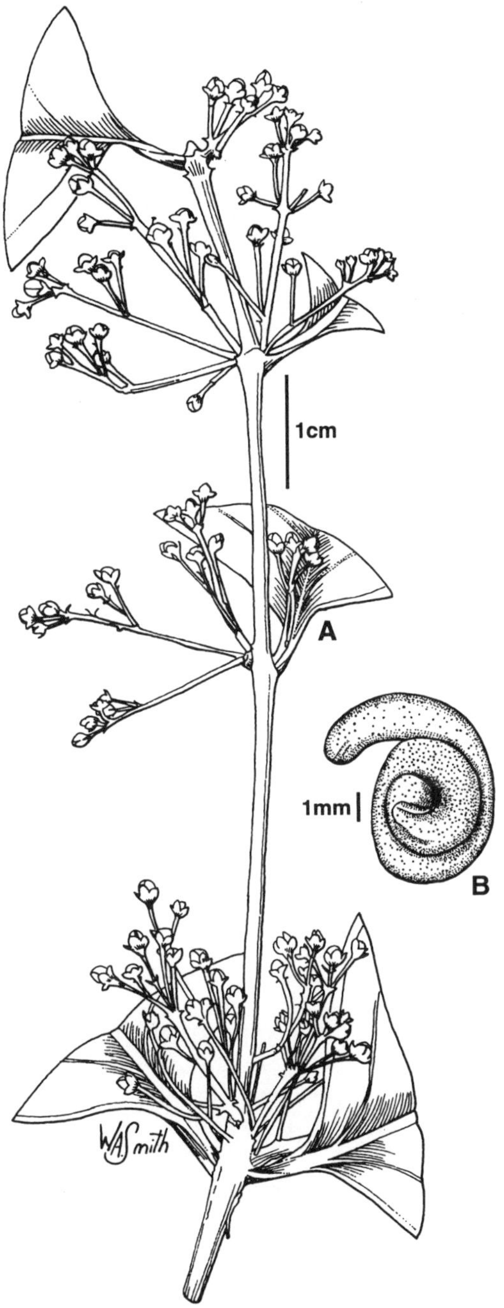

FIG. 10. *Gossia bidwillii*. A. Flowering branch. B. Embryo. (Based on: A, *Guymer 1125*; B, *Foster 13701*.)

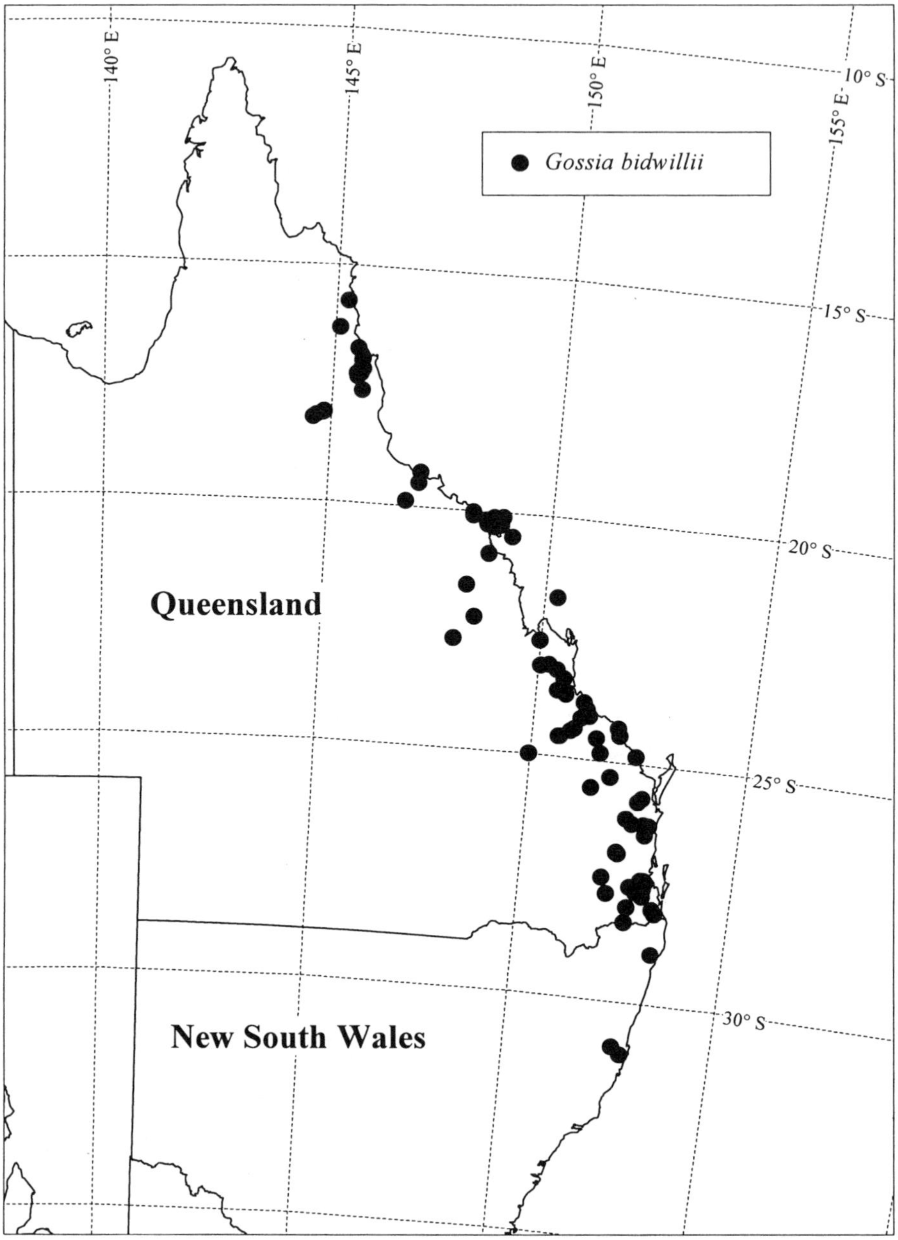

FIG. 11. Distribution of *Gossia bidwillii*.

PIF2859 (BRI, NSW); 4.5 km S of Binjour, Binjour Plateau, Numdubbera, 25°34′S, 151°27′E, *Forster PIF3323* (BRI, CANB); Coxens Peak, 22°12′S, 148°27′E, *Forster PIF7317* (BRI, MEL); Anvil Peak, Hodgson Ra., 22°41′S, 148°01′E, *Forster PIF7226* (BRI, MEL); Moores Ck, 23°20′S, 150°35′E, *Hyland 2334* (QRS); Carborough Ra., 1 km NW of Lake Elphinstone outlet, 21°33′S, 148°13′E, *Telford 11139 & Rudd* (CANB); Duaringa, Nov 1943, *White 12455* (BRI).—PORT CURTIS DISTRICT: Mt. Middle Percy Is., 55 km NE of Arthur Point, Shoalwater Bay, 21°40′S, 150°16′E, *Batianoff et al. 11664* (BRI); SF 114, Site 1 near Fairview Mt, 23°06′S, 150°02′E, *Batianoff 9812145* (BRI); S tributary of Dry Ck, Portion 3 Clifford, 34 km E of Biloela, 24°25′S, 150°51′E, *Brushe 17 & Hoy* (BRI); Rosedale, *Dovey 81, 157* (BRI); 6 km W of Canoona, 23°04′S, 150°13′E, *Forster PIF3388* (BRI, NSW); 1 km W of Targinie, Mt Larcom, 23°48′S, 151°06′E, *Forster PIF4729 & Batianoff* (BRI, CANB, MO, MEL); Crocodile Ck, Bouldercombe Gorge, 23°36′S, 150°28′E, *Forster PIF4992 & Hoy* (BRI, NSW); Dalma, 24°08′S, 151°04′E, *Francis 154* (BRI); Resumption LA, SF 391 Bulburin, 24°32′S, 151°28′E, *Forster PIF14523 et al.* (BRI); Roundabout Scrub, Dry Ck, SF 316 Kroombit, 24°22′14″S, 150°56′18″E, *Forster PIF20144* (BRI); Dan Dan Scrub, Calliope Shire, 24°S, 151°E, *Gibson 159* (BRI); Deepwater NP, 40 km E of Miriam Vale, 24°17′S, 151°56′E, *Gibson TOI897* (BRI); Holligans Ck, ca. 20 km S of Rockhampton, 23°40′S, 150°40′E, *Hemsley 7109* (CANB, NSW); Glenprarie, ca. 30 km N of Marlborough, 22°35′S, 149°58′E, *Hemsley 7117* (CANB, NSW); beside Mt Archer Rd, 23°21′S, 150°35′E, *Hoy 141* (BRI); Mt Etna, 23°10′S, 150°25′E, *Hyland 4341RFK* (BRI); SF 150, 4 km NNW of Calliope, 23°57′S, 151°11′E, *Worthington 1456* (BRI).—BURNETT DISTRICT: Goodnight Scrub, SF 169 St. Agnes, Kalliwa Ck, 25°19′S, 151°52′E, *Forster PIF13882* (BRI); Mt Perry, [collector illegible], (BRI-93350).—WIDE BAY DISTRICT: SF 234, SW of Cooroy, 26°28′S, 152°49′E, *Bean 7141* (BRI, NSW); Oakview SF 220, 4.7 km past barracks, 26°09′S, 152°20′E, *Forster PIF4879 & Bird* (BRI, MO, NSW); Tinana Ck, 7 km E of Tiaro, 25°42′S, 152°39′E, *Forster PIF12515 & Smyrell* (BRI); Graphite Mine Rd, Guyru Ra., Mt Bauple area, 25°47′S, 152°34′E, *Forster PIF12530 & Smyrell* (BRI); Mt Gaeta, ca. 36 km N of Mt Perry township, 24°50′S, 151°35′E, *Forster PIF14151* (BRI, MEL, NSW); Graphite Mine Rd, Mt Bauple, 25°47′S, 152°34′E, *Forster PIF14625* (BRI, MEL); property of N. Dargusch, Cattermul Avenue, Burnet Shire, 24°50′43″S, 152°24′37″E, *Forster PIF20203* (BRI); "Green Horizons," property of A. Geckeler, 8.4 km from Rules Beach Rd along Fingerfield Rd, ca. 60 km direct line NW of Bundaberg, 24°26′S, 151°59′E, *Geckeler et al. 138* (BRI, CANB); Kin Kin, 26°16′S, 152°53′E, *Priest s.n.* (BRI AQ457914); Lone Pine Farm, Gympie, 26°S, 152°E, *Priest s.n.* (BRI AQ459445); Mt Glastonbury, ca. 20 km W of Gympie, 26°15′S, 152°29′E, *Sharpe 4624 & Forster* (BRI, NSW).—DARLING DOWNS DISTRICT: Cawdor near Toowoomba, 27°2–0′S, 151°E, *Gilbert 25* (BRI); Killarney, *Wedd 1106* (BRI); Bunya Mts, Oct 1919, *White* s.n. (BRI AQ93341).—MORETON DISTRICT: near White Swamp Rd, SSW of Boonah, 28°19′S, 152°34′E, *Bean 1375* (BRI); 11.3 km from Moore along road to Blackbutt, ca. 400 m along forestry rd to S just before Blackbutt Ck, 26°52′S, 152°12′E, *Beesley 981 & Ollerenshaw* (BRI, CANB); Woogaroo Ck, Goodna, 27°S, 152°E, Nov 1980, *Bird s.n.* (BRI AQ344784); Woogaroo Ck, Goodna, Ipswich, 27°S, 152°E, Oct 1983, *Bird & Smith s.n.* (BRI AQ431242); Upper Brookfield, near Brisbane, *Blake 15469* (BRI); Laheys Lookout, Lamington NP, 28°00′S, 153°11′E, Nov 1986, *Davidson s.n.* (BRI AQ407732); Commissioner's View, Blackbutt Ra., Nanango, 26°52′S, 152°12′E, *Forster PIF3248 et al.* (BRI, CANB); Falls Ck, 4 km NW of W Haldon, 27°45′S, 152°04′E, *Forster PIF4758 & Bird* (BRI); SF 607 Dinden, Bridle LA, 16°58′S, 145°36′E, *Forster PIF14467* (BRI, MEL); Enoggera Scientific Area, SF 309, Scrub Rd, D'Aguilar Ra., 27°25′22″S, 152°50′24″E, *Grimshaw G320 & Williams* (BRI); Enoggera Ck, Brisbane Forest Park, 50 m below Scrub Rd, 27°25′S, 152°50′E, *Halford Q1829* (BRI); Canungra, property of K. Curtis, 28°S, 153°E, *Hockings 2* (BRI); Moggill SF, SW of Brisbane, 27°S, 152°E, *Jessup 267* (BRI, CANB, NSW); Mt Crosby area between Mt Crosby and Anstead, between water pipeline and bank of Brisbane R., 27°S, 153°E, *Jessup 269* (BISH, BRI, NSW); Moggill SF, SW of Brisbane, 27°S, 152°E, *Jessup 274* (BRI); Benarkin SF (283 Colinton), E side of Balfur Ra. on D'Aguilar Hwy, 26°53′S, 152°12′E, *McDonald 4130 & Williams* (BRI); just S of D'Aguilar Ra. Hwy, Balfour Ra. (Blackbutt Ra.) E of Benarkin, 26°53′S, 152°12′E, *McDonald 4152* (BRI, NSW), *McDonald 4153* (BRI, MO); Sarabah Ra., ca. 10 mi S of Canungra, *Schodde 5581* (BRI, CANB); Blackbutt Ra., 8 km W of Moore, 26°54′S, 152°14′E, *Williams 75123* (BRI); Marburg, 27°34′S, 152°36′E, *Williams 83006* (BRI); near Boonah, Charlwood, NW slopes of Mt French, 28°00′S, 152°35′E, *Williams 83010* (BRI, NSW); Woogaroo Ck, S of Goodna, 27°40′S, 152°45′E, *Williams 83028* (BRI, MO, NSW).—NEW SOUTH WALES: Lismore, *Bauerlen 517* (MO); Ballina, *Bauerlen 582* (NSW); Tintenbar, *Bauerlen 584* (BRI); Dunoon, *Bauerlen 1411* (NSW); Hickeys Ck, Macleay R., 30°56′S, 152°37′E, *Fitzgerald s.n.* (MEL 60621); Pelican Ck Reserve, Ruthven, 28°56′S, 153°17′E, *Floyd 2007* (QRS); Nymboida River, Eastern Dorrigo, *Hayes NSW144729* (NSW); Casino, *Irby NSW 180698* (NSW); Macleay R., 31°05′S, 152°50′E, *Macdonald 28* (MEL); Whian Whian SF, W of Mullumbumby, *Webb & Tracey 8146* (BRI); Dorrigo, *Welch NSW144730* (NSW).

Gossia bidwillii most resembles *G. acmenoides* (no. 9) from which it differs by its usually deeply channeled petioles and much more prominent oil glands in the leaves. It also tends to grow in drier sites than *G. acmenoides*. It can also be confused with *G. myrsinocarpa*, which has 4-merous flowers and greenish (vs. whitish) sepal lobes. *Gossia shepherdii* differs in its flat petioles and far fewer oil glands on the adaxial leaf surface. Seedlings of *G. bidwillii* have branchlets with four wings (*Gray 2319*, QRS; *Hyland 11637*, QRS).

Gossia bidwillii is the most widespread Australian species in the genus and exhibits considerable variation throughout its range, which is typical for widespread species. We concur with Smith (1956: 36) that *Myrtus racemulosa* is a synonym of *G. bidwillii*, since the putative differences recognized by Bentham (1867) are inconsistent. Regional variants of *G. bidwillii* may warrant additional investigation as possibly recognizable taxa. Features of the bark, in particular, may be more diagnostic than we were able to ascertain. Northern populations reportedly with greenish bark (*Austromyrtus* sp. "Danbulla, L. S. Smith 10123" [Henderson 1997: 125; Hyland et al. 1999]) merit further investigation in the field (see herbarium notations on: *Guymer 2071*, BRI; *Moriarty 1155*, BRI; *Forster PIF14447*, BRI; *O'Farrell* 84, BRI, QRS; *Gray 2319*, BRI; *Gray 2781*, BRI); A. Ford (pers. comm.) indicates that these northern populations maintain the greenish bark throughout the year, even though other populations lack the greenish hue. Caution should be exercised, since bark in Myrtaceae is well known for its ability to change color and appearance during the year.

Fensham (1995a, b) reported on the abundance of "*Austromyrtus* sp. Forty Mile Scrub *G. C. Stocker 1785*" (Henderson 1997; Hyland et al. 1999) in the dry rainforests in north Queensland, but was unable to find substantive differences between it, "typical" specimens of *G. bidwillii*, and the previously recognized *Austromyrtus racemulosa* (Fensham, pers. comm.). A similar view is held by W. J. F. McDonald (pers. comm.), who has seen *G. bidwillii* throughout much of its range. In contrast, P. Forster (pers. comm.) believes the Forty Mile populations are generally more shrublike and have a greater degree of branching. In upland, relatively dry areas, such as the Forty Mile Scrub region, populations with individuals of reduced stature and with greater branching are not unexpected from a physiological perspective. Evolutionary theory predicts exactly this sort of regional variation in widespread species.

Two collections (*Forster PIF13701* and *PIF14294*), tentatively designated earlier as the undescribed species *Austromyrtus* sp. "Spencer Creek" (Henderson 1997), tend to have less prominent oil glands on the lower leaf surface, greenish (as opposed to whitish) sepals, petals slightly ciliate near the apex of the upper surface, and fewer stamens (70–90). Given the widespread variation in *G. bidwillii* and our use of a species criterion that requires fixed morphological differences as a minimum condition of specific recognition, these populations do not warrant recognition at the specific level.

Some photos in Williams's account of native plants of Queensland (1984: 48) are incorrectly labeled. A flowering specimen of *G. bidwillii* is identified as *Austromuellera trinervia* (Proteaceae). The two photos at the bottom (flowering branch and trunk) identified as "*Austromyrtus bidwillii*" appear to be *G. acmenoides*. In addition, *G. bidwillii* and *G. acmenoides* cannot be confidently separated on the basis of bark, which appears to be a misconception among some workers (e.g., Floyd 1989: 243–245). As noted by Williams (1984), given the beauty of the bark, it is surprising that *G. bidwillii* and *G. acmenoides* have not become more popular ornamentals.

The species is named in honor of John Carne Bidwill (1815–1853), who collected the holotype.

9. Gossia acmenoides (F. Mueller) N. Snow & Guymer, comb. nov. *Myrtus acmenoides* F. Mueller, Fragm. 1: 77. 1859. *Austromyrtus acmenoides* (F. Mueller) Burret, Notizbl. Bot. Gart. Berlin-Dahlem 15: 501. 1941.—TYPE: AUSTRALIA. Queensland. Moreton District: Wide Bay, *Moore s.n.* (lectotype, here designated: MEL!).

Austromyrtus acutiuscula Burret, Notizbl. Bot. Gart. Berlin-Dahlem 15: 502. 1941.—TYPE: AUSTRALIA. New South Wales: Murwillumbah, Nov 1892, *W. Bauerlen s.n.* (isotype: NSW!).

Multi-stemmed shrubs to trees, 2.5–18 m tall. Bark of trunk smooth, shedding in small flakes, whitish or brown to reddish. Branchlets rounded, not winged; light brown to gray; bark smooth, glabrous; oil glands few to common, indistinct to prominent. Leaves mostly slightly discolorous, matte to somewhat glossy above, matte to slightly glossy below. Stipules consisting of relatively stout ferrugineous setose hairs, often obscure. Petioles 1.5–4 mm long, slightly gray to black, smooth above to sometimes slightly channeled (rarely deeply so), eglandular. Leaf blades 2.0–7.5 (–11.0) cm long, 1.0–4.0 (–6.0) cm wide, narrowly elliptic to broadly elliptic; base cuneate, apex acute to acuminate, the tip rounded or pointed; margins revolute; adaxial surface mostly glabrous, oil glands common to abundant (viewed with magnification), midvein flush to raised; abaxial surface glabrous, oil glands sparse to common, visible with magnification; secondary veins indistinct to prominent; intramarginal veins indistinct to prominent, the inner intramarginal vein 1–3 mm from edge. Inflorescence terminal or axillary, a single flower or panicle, solitary, paired, or fascicled, anthopodia absent, metaxyphylls absent. Peduncles 2–15 mm long, rigid, glabrous. Bracteoles 2, 0.7–1.7 mm long, 0.2–0.5 mm wide, linear to narrowly triangular, rigid, scalelike, midrib absent, the tips not exceeding base of calyx lobes, glabrous to sparsely sericeous, mostly persistent in fruit. Hypanthium 1.5–2.5 mm long, cupulate to obconic, glabrous to sericeous, oil glands common. Calyx lobes 5, 1.5–2 mm long, broadly ovate to rounded, apex obtuse, greenish; adaxial surface shortly and sparsely sericeous; abaxial surface sparsely sericeous. Petals 5, 3–5 mm long, 3–4 mm wide, white or pinkish, widely ovate to oblate; adaxial surface mostly glabrous or sometimes sericeous on edges; abaxial surface glabrous, oil glands few to common, sometimes bright yellow. Stamens 100–120, included, filaments (2.5–) 5–6 mm long; staminal disk glabrous; anthers 0.5–0.7 mm long, subglobose to globose, the tip of connective with a single gland. Style 2.5–4 mm long, glabrous to sparsely sericeous; ovary apex slightly hairy; ovules 8–10 per placenta, placed irregularly. Berry 4–6.5 mm long, 5.5–7 mm in diameter, subglobose, base rounded, glabrous, dark purple to black. Seeds in mature fruit 1–4; nearly round to slightly reniform, smooth, ivory-colored, micropylar operculum present, seed coat hard, bony. Embryo C-shaped to circinate, starchy, oil glands absent, endosperm absent; hypocotyl relatively thin, tip near radicle not or barely swollen, tip at same plane as embryo; cotyledons mostly straight, shorter than hypocotyl, folded back towards hypocotyl.

Phenology. Flowering in October and November; fruiting December through April.

Distribution (Fig. 12). From east-central New South Wales to central Queensland; locally common to very common in subtropical rainforests and vineforests from ridgetops to alluvial flats, often on rich substrates, such as those from Tertiary basalts; 50–600 m.

Vernacular name. Scrub ironwood.

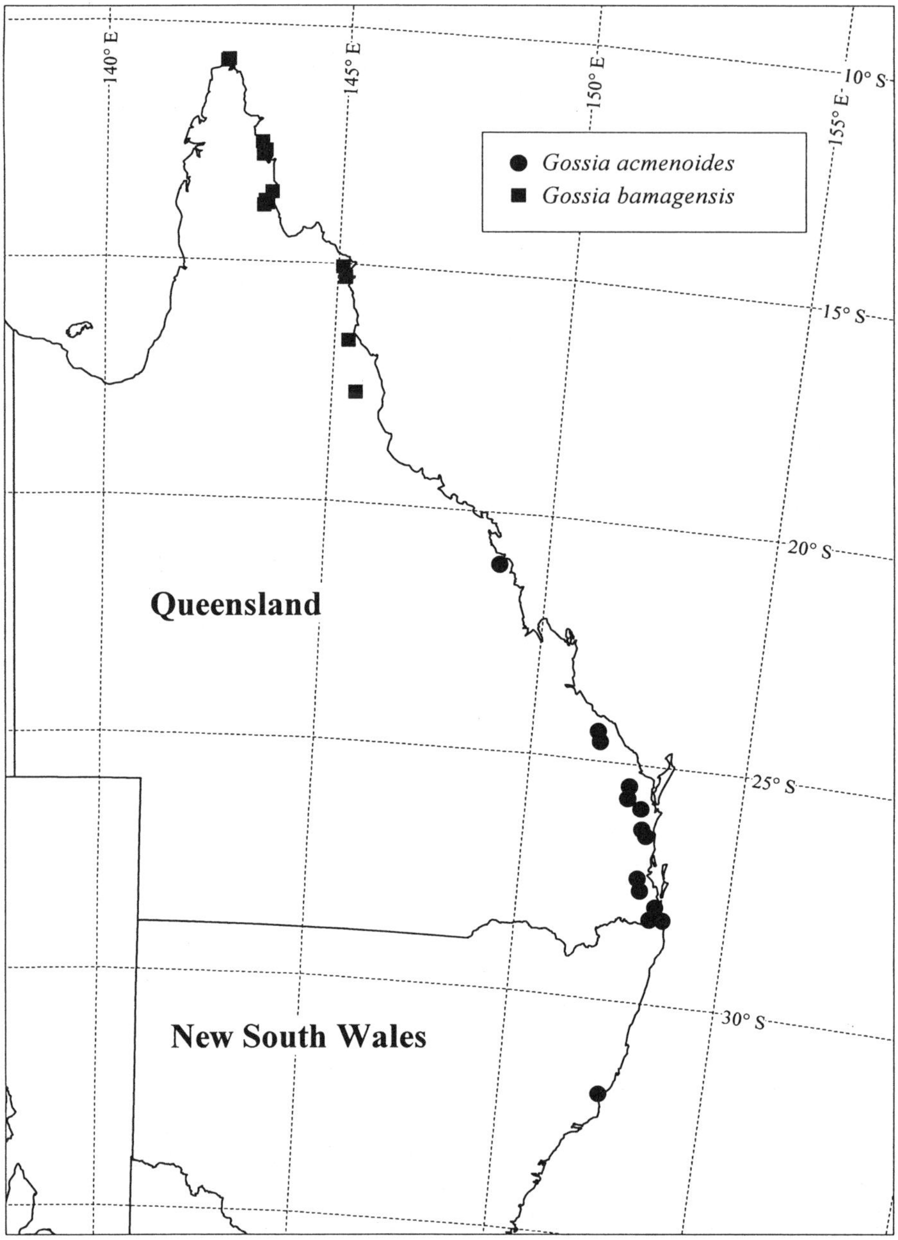

FIG. 12. Distribution of *Gossia acmenoides* and *G. bamagensis*.

ADDITIONAL SPECIMENS EXAMINED. **Australia.** QUEENSLAND. SOUTH KENNEDY DISTRICT: Eungella South, 21°S, 148°E, *Smith 4787* (BRI).—PORT CURTIS DISTRICT: Bulburin SF (=Granite Ck SF), 24°S, 151°E, *McDonald 3436* (BRI, CANB); Resumption LA, SF 391, Bulburin, 24°32′S, 151°28′E, *Forster PIF14519 et al.* (BRI, MEL); Boyne LA, SF 391 Bulburin, 0.5 km along Scott Rd off Boyne Rd, 24°32′S, 151°29′E, *Forster PIF14563 et al.* (BRI); Colosseum Ck TR 115, Many Peaks Ra., 24°23′S, 151°27′E, *Forster PIF14636* (BRI).—WIDE BAY DISTRICT: Home LA, SF 502, N of Gympie, 26°05′S, 152°43′E, *Bean 6683* (BRI); Glenwood, property of J. Bolderrow, 25°57′S, 152°35′E, *Forster PIF12475* (BRI); Fairlies Knob area, Compartment, Doongul LA, SF 1294, Warrah, 25°29′S, 152°17′E, *Forster PIF12573 & Smyrell* (BRI, MEL); Stoney Ra., SF 38, 25°37′S, 152°19′E, *Forster PIF13307 & Smyrell* (BRI); west side of Grassy Mt, SSW of Bauple, 25°51′S, 152°35′E, *Guymer 2082* (BRI); Imbil, 26°27′S, 152°40′E, *Guymer 2089* (BRI); Mt Cooroy, S of Tewantin-Cooroy Rd, northern slope, 26°S, 152°E, *Jessup 208 & Sharpe* (BRI); N of Mt Eewah, close to Eumundi-Kenilworth Rd, 26°S, 152°E, *Jessup 218 & Sharpe* (BRI); Gympie, 1906, *Kenny s.n.* (BRI AQ093324); Amamoor, 26°21′S, 152°41′E, *Smith 5382* (BRI); Kin Kin, Jan 1917, *White s.n.* (BRI 93326).—MORETON DISTRICT: Emundi, Nov 1894, *Bailey & Simmonds s.n.* (BRI 93442); Brisbane R., *Dietrich s.n.* (MO, SING); near end of Grandview Rd, Moggill, 27°34′S, 152°53′E, Jan 1988, *Bird s.n.* (BRI AQ436256); Brocks Rd near Nicholls Scrub, Currumbin Valley, 28°11′24″S, 153°25′38″E, *Grimshaw G162* (BRI); old dairy paddock beside road to Egg Rock, Lamington NP, 28°11′S, 153°11′E, *Grimshaw G176N* (BRI); Kiamba–Nambour Rd near Wappa Dam, 26°33′51″S, 152°54′56″E, *Grimshaw G201* (BRI); Guanaba Valley, E of Mt Tamborine, 27°S, 153°E, Jul 1986, *Hauser s.n.* (BRI AQ440296); Wards Scrub, W of Samford near headwaters of South Pine R., 27°22′S, 152°48′E, *Jessup 572 & Guymer* (BRI, NSW); Tonys Ck, Numinbah Prison Reserve, 28°08′S, 153°12′E, *McDonald 6214* (BRI); The Summit, Mt Glorious, at the end of Alex Rd, 27°S, 152°E, Nov 1980, *Moriarty s.n.* (BRI AQ344964, CANB); The Cougals, Springbrook NP, Mt Cougal section, 28°14′S, 153°19′E, *Forster PIF15832 & Leiper* (BRI, L); Moreton Bay, *Mueller s.n.* (MEL 92353); Moggill SF, near Grandview Rd, 27°S, 152°E, *Orford s.n.* (BRI AQ564624), Moggill SF, NE corner, 0.1 km along first turnoff in Forest from Mill Rd, 0.8 km from intersection with Grandview Rd, 27°32′S, 152°50′E, *Snow & Snow 7741* (BRI, GREE); 609 Grandview Rd, property of G. Nosworthy, Pullenvale, ca. 3.5 km from turnoff on Moggill Rd, 27°32′S, 152°50′E, *Snow 7745* (BRI, GREE).—NEW SOUTH WALES: Mullumbimby, May 1895, *Bauerlen s.n.* (NSW 144697); Tumbulgum, Dec 1897, *Bauerlen s.n.* (NSW 144696); Walcha, Feb 1885, *Crawford s.n.* (MEL 60615); Acacia Ck, *Dunn 231* (NSW); Boonabilla Ck, 3 km S of Blue Gum Flat, *Floyd 1941* (BRI); Pelican Ck, Ruthven, *Floyd 2012* (BRI, MEL, NSW); S branch of Hopping Dicks Ck, NW of Limpinwod, 28°18′S, 153°11′E, *Guymer 1805 & Jessup* (BRI); Nymboida R., May 1958, *Hayes s.n.* (NSW 144701); Woodburn, Nov 1903, *Maiden & Boorman s.n.* (NSW 144698); Avondale, S [of] Wollongong, Nov 1983, *Mills s.n.* (NSW 180702); Lismore, Dec 1906, *Rothwell s.n.* (NSW 144704); Nirrim ex Mebbin SF (cult.), 31°55′S, 152°31′E, Nov 1990, *Walker s.n.* (BRI AQ500937); Lynch's Ck, Mar 1944, *White 12551* (BRI); Clarence R., *Wilcox s.n.* (NSW 144699, NSW 144700).

The smooth bark, which is generally a patchwork of orange and brown, is partially diagnostic of *G. acmenoides*, but this feature is shared with the more widespread species *G. bidwillii*. The latter has much more pronounced oil glands on the leaves (usually visible without magnification), a more deeply channeled upper petiole, and leaves that emit a much stronger smell and stick much more readily to the hands when crushed. The range of *G. acmenoides* overlaps with that of *G. bidwillii* in southeastern Queensland and northern New South Wales, but the latter tends to grow in drier sites.

The leaves of *G. acmenoides* can vary considerably in size, from as short as 2 cm to as long as 11 cm. No positive correlation exists between leaf and inflorescence size, e.g., *Simmonds s.n.* (BRI AQ093427) has large leaves but short inflorescences. The handsome bark is striking, suggesting this species deserves wider horticultural use.

The specific epithet *acmenoides* alludes to a resemblance to species of *Acmena* (Myrtaceae).

10. Gossia gonoclada (F. Mueller ex Bentham) N. Snow & Guymer, comb. nov. *Myrtus gonoclada* F. Mueller ex Bentham, Fl. austral. 3: 275. 1867. *Austromyrtus gonoclada* (F. Mueller ex Bentham) Burret, Notizbl. Bot. Gart. Berlin-Dahlem 15:

501. 1941.—TYPE: AUSTRALIA. Queensland: Moreton District: "Moreton Bay" (Mogill), *C. Stuart s.n.* (=385 at MEL) (holotype: K!, photo: BRI!; isotype: BRI! MEL!).

Trees, 6–18 m tall. Bark of trunk smooth to flaky, brown, reddish, orangish, or gray. Branchlets distinctly 4-angled to laterally compressed, edges distinctly winged, light brown to reddish brown, bark smooth or slightly flaky, glabrous to sparsely sericeous or villous, oil glands few to common but indistinct. Leaves slightly discolorous, matte or slightly glossy above, matte below. Stipules type A or type B (see Morphology). Petioles 2–4 mm long, light brown to reddish brown, sometimes channeled slightly near base, eglandular. Leaf blades 2.5–6.0 cm long, 1.0–3.7 cm wide, elliptic to obovate (occasionally narrowly so); base cuneate; apex obtuse to mostly acute, the tip mostly rounded to pointed, rarely mucronate, slightly retuse, or falcate; margins flat; adaxial surface glabrous; oil glands abundant, prominent (viewed with magnification); midvein slightly raised (particularly proximally); abaxial surface glabrous, oil glands abundant, prominent (viewed with magnification); secondary veins indistinct to prominent; intramarginal veins indistinct to prominent, the inner intramarginal vein 1–4 mm from edge. Inflorescence terminal or axillary, solitary; a single flower or a raceme; anthopodia <2 mm long or absent, metaxyphylls absent. Peduncles 6–15 mm long, rigid, sericeous to sparsely sericeous. Bracteoles 2 (usually falling in bud), 1.8–2 mm long, 0.5–0.8 mm wide, linear to narrowly triangular, rigid, scalelike without midrib, with tips shorter than base of calyx lobes, sparsely sericeous, persistent in fruit. Hypanthium 2–3 mm long, obconic to cupulate, sericeous to villous, oil glands common to abundant. Calyx lobes 5 (rarely 4), 1–2 mm long, broadly triangular, apex obtuse or somewhat acute, greenish; adaxial surface sparsely sericeous to densely sericeous; abaxial surface sparsely sericeous to sericeous. Petals 5, 3–4 mm long, 2.5–3.5 mm wide, widely ovate, white; adaxial surface glabrous and somewhat ciliate along lower margins; abaxial surface glabrous, oil glands common to abundant. Stamens up to 130, included, filaments up to 5 mm long; staminal disk mostly glabrous or sparsely short-villous; anthers 0.4–0.6 mm long, subglobose, the tip of connective with a single gland. Style 3.5–5 mm long, glabrous or sparsely sericeous; ovary apex glabrous to hairy; ovules 20–30 per placenta, placed irregularly. Berry 8–10 mm long, 7–8.5 mm in diameter, subcylindrical, base tapered, glabrous, dark blue-black. Seeds in mature fruit 1–3, circular to subglobose, somewhat flattened, smooth, ivory-colored, micropylar operculum present, seed coat hard and bony. Embryo C-shaped to circinate, starchy, oil glands absent, endosperm absent; hypocotyl relatively thick, tip near radicle swollen, tip at same plane as embryo; cotyledons mostly straight, shorter than hypocotyl, pointing forward.

Phenology. Flowering in November and December; fruiting January through early April.

Distribution (Fig. 13). Southeastern Queensland in the McPherson-Macleay area of endemism along the lower reaches of the Brisbane and Logan rivers; in remnant riverine rainforests on river banks, terraces, and slopes; elevation 5–50 m.

Conservation status. The species is considered Critically Endangered (criteria D, Species Survival Commission 1994). Conservationists in the greater Brisbane metropolitan area are actively working to safeguard the preservation of this species.

Vernacular name. Angle-stemmed myrtle.

ADDITIONAL SPECIMENS EXAMINED. **Australia.** QUEENSLAND. PORT CURTIS DISTRICT: Keppel Bay, 1872, *Eaves s.n.* (MEL 72380).—MORETON DISTRICT: Brisbane, *Bailey s.n.* (BRI AQ93441); ca. 1 km along Sevelle

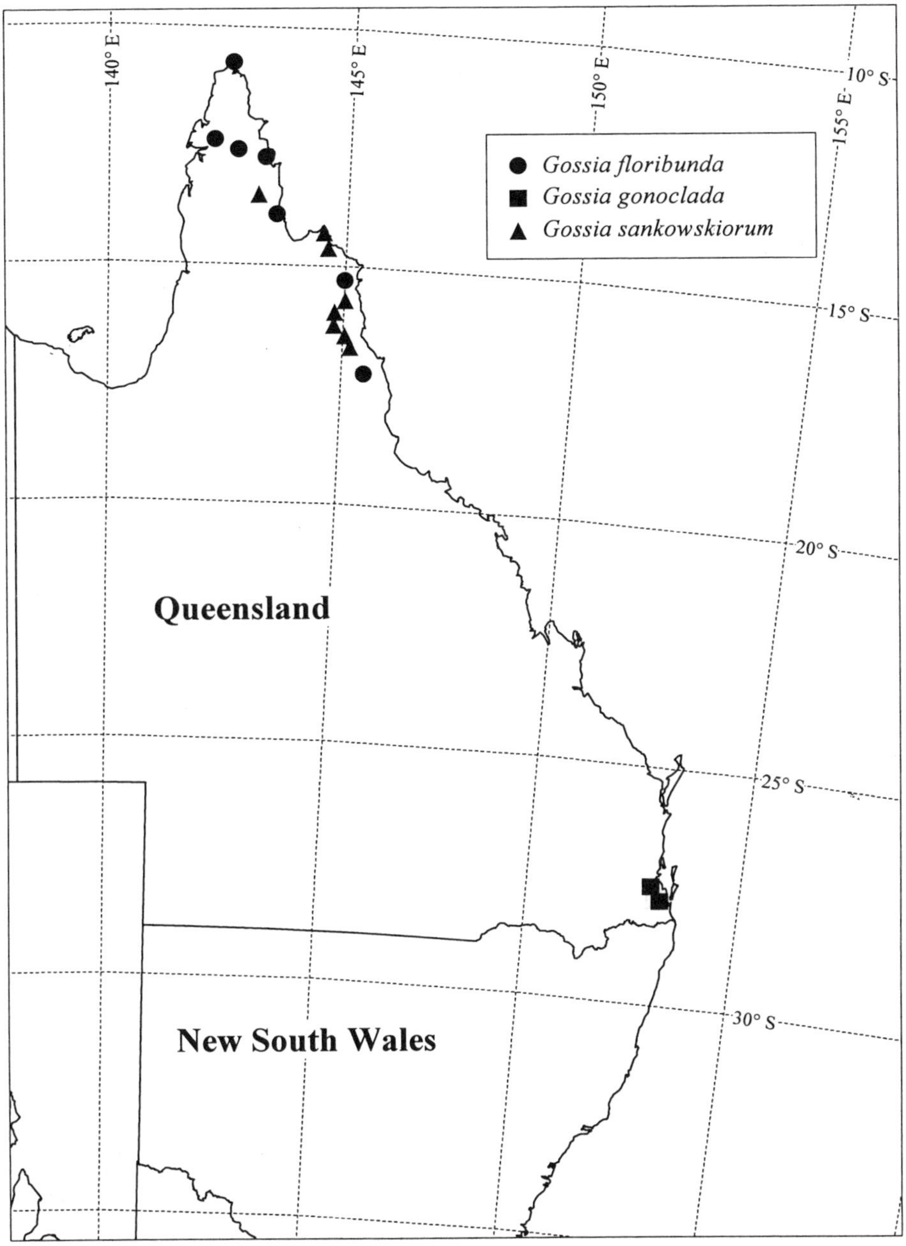

FIG. 13. Distribution of *Gossia floribunda*, *G. gonoclada*, and *G. sankowskiorum*.

Rd from the Pacific Hwy, ca. 7 km NW of Beenleigh, 27°40′S, 153°09′E, *Beesley 875 & Ollerenshaw* (BISH, BRI, CANB, NSW); Brisbane R., 1863–1865, *Dietrich 209* (BRI AQ93444, CANB); Murray's Rd, Loganholm, 27°40′S, 153°09′E, *Forster PIF15113 & Leiper* (BRI); Nosworthy Park, 27°32′S, 152°59′E, Aug 1991, *Grimshaw s.n.* (BRI AQ546691); Murray's Rd beside Slacks Ck, Tanah Mera, 27°39′52″, 153°09′37″E, *Grimshaw G342 & Williams* (BRI); Cliveden Ave, Lots 3 and 4, Corinda, 27°33′S, 152°59′E, *Halford Q2877 et al.* (BRI); Cliveden Ave, Lot 2, 27°33′S, 152°59′E, *Halford Q2879* (BRI); Murray's Rd, Loganholm, 27°39′S,1

53°09′E, *Hauser s.n.* (BRI AQ424470), Dec 1986, *Leiper s.n.* (BRI 414081), Mar 1986, *Leiper s.n.* (BRI AQ431237), Feb 1988, *Leiper s.n.* (BRI AQ437294, NSW); Slacks Ck, Usher Park, Daisy Hill, 27°38′S, 153°09′E, Dec 1990, *Leiper s.n.* (BRI AQ501219); Slacks Ck, Murray's Rd, Loganholm, 27°40′S, 153°09′E, Nov 1991, *Leiper s.n.* (BRI AQ509451); boundary fence of Alexander Clark Park & Jensen Tree Farm at Loganholm, 27°42′S, 153°11′E, Jun 1995, *Leiper s.n.* (BRI AQ637289); along bank of Slacks Ck, Tanah Merah, 27°39′S, 153°09′E, Feb 1996, *McIlay s.n.* (BRI AQ584934); Nosworthy Park, 27°32′S, 152°59′E, May 1993, *Playford s.n.* (BRI AQ569437, MEL 717893); Nosworthy Park, Corinda, 27°32′S, 152°59′E, *Snow & Guymer 7740* (BRI, GREE); Murray's Rd, Slacks Ck, Brisbane, 27°39′S, 153°09′E, *Thomas RFR39 & Barry* (BRI); Slacks Ck/Murray Rd, Beenleigh, 27°40′S, 153°09′E, *Thomas 309 & Barry* (BRI); Nosworthy Park, Corinda, Brisbane, 27°32′S, 152°59′E, Jan 1993, *Thomas s.n. & Barry* (BRI AQ636361); Usher Park, soccer and rugby fields, Daisy Hill, 27°38′S, 153°09′E, *Thomas RFR14 & Gale* (BRI).

The winged branchlets, slightly retuse leaf apex (check closely), and 5-merous flowers distinguish *G. gonoclada* from congeners in southeastern Queensland.

Shapcott and Playford (1996) demonstrated relatively high levels of genetic diversity in *G. gonoclada* (as *Austromyrtus gonoclada*) relative to other species now assigned to *Gossia*, despite its currently small population size of fewer than 100 individuals. Individual trees often flower only once in two or more years.

The specific epithet derives from the Greek *goni* (angled) and *clados* (branch or shoot), in reference to the two narrow wings on each side of the round to somewhat flattened internodes of branchlets.

11. Gossia hillii (Bentham) N. Snow & Guymer, comb. nov. *Myrtus hillii* Bentham, Fl. austral. 3: 275. 1867. *Austromyrtus hillii* (Bentham) Burret, Notizbl. Bot. Gart. Berlin-Dahlem 15: 501. 1941.—TYPE: AUSTRALIA. Queensland: Moreton District, Moreton Bay, *E. F. A. Fitzalan s.n.* (lectotype, here designated: K!, photo: BRI!; isolectotype: MEL!).

Myrtus opaca C. T. White, Proc. Roy. Soc. Queensland 53: 218. 1942. *Austromyrtus opaca* (C. T.White) L. S. Smith, Proc. Roy. Soc. Queensland 67: 35. 1956.—TYPE: AUSTRALIA. Queensland, *C. T. White 9592* (holotype: BRI!).

Shrubs to trees, 2–12 m tall. Bark of trunk rough, flaky, fissured, or somewhat tessellated, gray to dark gray. Branchlets rounded, not winged, light brown, smooth but soon becoming flaky, glabrous to sparsely sericeous, oil glands common but indistinct. Leaves more or less discolorous, mostly very glossy above, matte below. Stipules type A or type B (see Morphology). Petioles 1–7 mm long, light to dark brown, slightly to strongly channeled, eglandular. Leaf blades 2.0–6.2 cm long, 1.2–3.3 cm wide, narrowly elliptic to elliptic, base cuneate, apex acute to mostly acuminate, the tip rounded; margins flat; adaxial surface sericeous on lower portion of midvein but becoming glabrous, oil glands not visible from above, midvein raised prominently; abaxial surface glabrous to sparsely sericeous at base on midvein; oil glands few to common, prominent (viewed with magnification); secondary veins indistinct to not visible to the naked eye (usually more prominent above); intramarginal veins often not visible without magnification, the inner intramarginal vein 1–3 mm from edge. Inflorescence terminal or axillary, solitary; flowers mostly single or sometimes in triads; anthopodia and metaxyphylls absent. Peduncles 5–24 mm long, rigid, glabrous. Bracteoles 2, 1–3 mm long, 0.3–0.4 mm wide, linear, firm, scalelike without midrib, with tips shorter than base of calyx lobes, sparsely sericeous, persistent or mostly caducous in fruit. Hypanthium 2–2.5 mm long, cupulate, sericeous to densely sericeous throughout, oil glands common to abundant. Calyx lobes 5, 1.5–2.8 mm long, broadly triangular, apex obtuse to nearly acute, greenish; adaxial surface densely villous; abaxial surface glabrescent below. Petals 5, 5–6 mm long, 3.5–4.5 mm wide, elliptic to widely obovate, white; both surfaces shortly hairy throughout, sometimes densely so; oil glands few (faint and obscured by hairs). Stamens 120–135, included, filaments 3.5–4 mm long; staminal disk short-villous (sometimes densely so); anthers 0.5–0.7 mm long, subglobose to globose, the tip of connective with a single gland. Style ca. 4.5 mm long, sparsely sericeous, ovary apex often hairy near base of style; ovules up to 16 per placenta, around edge in a single row or placed irregularly. Berry 7–11 mm long and in diameter, subglobose to globose, base rounded, glabrous to sparsely sericeous, dark blue-black. Seeds in mature fruit 1–3, circular to subglobose and somewhat flattened, smooth, light brown, micropylar operculum present or absent, seed coat hard, bony. Embryo circinate, occasionally C-shaped, starchy or slightly oily, oil glands absent, endosperm absent; hypocotyl relatively thick (cotyledons very thin), tip near radicle not or barely swollen, tip at same plane as embryo; cotyledons circinate (held tightly within center of circinate coil), shorter than hypocotyl, pointing forward.

Phenology. Flowering from October through December; fruiting November through April.

Distribution (Fig. 14). From northern New South Wales to southern Queensland, and disjunct in central and northeastern Queensland (Hyland et al. 1999); uncommon to locally common in subtropical rainforests and notophyll or microphyll vineforests, occasionally near edge of *Eucalyptus* woodlands and river flats; 20–1100 m.

Vernacular name. Scaly myrtle.

ADDITIONAL SPECIMENS EXAMINED. **Australia.** QUEENSLAND. COOK DISTRICT: Mt Misch, tributary of Oakey Ck, Herberton Ra., 17°14′S, 145°25′E, *Ford 2029* (QRS, GREE); SFR 185, Downfall LA, EP/2, 17°08′S, 145°35′E, *Risley 509* (QRS); Downfall LA, R. 185, 17°10′S, 145°37′E, *Sanderson 223, 224* (BRI, QRS); Exp. E/P2, Downfall LA, R. 185, 17°10′S, 145°37′E, *Sanderson 257* (BRI, QRS); EP 40, SFR 144, 16°16′S, 147°07′E, *Sanderson 1658* (QRS).—NORTH KENNEDY DISTRICT: "Taravale," Coane Ra., E of the Paluma–Taravale road, site 39, 19°05′S, 146°08′E, *Ford 3097& Holmes* (BRI, GREE).—SOUTH KENNEDY DISTRICT: Bee Ck, Sep–Oct 1947, *Clemens s.n.* (BRI 93433, MEL 517085); Dalrymple Heights and vicinity, Nov 1947, *Clemens s.n.* (BISH, BRI AQ93431, BRI AQ93440); North Rd, SF 652 Cauley (Cathu), ca. 1.5 km SSW of Bloomsbury, 20°47′S, 148°28′E, *McDonald 6031 & Champion* (BRI).—PORT CURTIS DISCTRICT: Granite Ck, ca. 250 m N of Scott's Rd crossing, Bulburin SF, 24°34′S, 151°31′E, *Thomas RFR137 et al.* (BRI).—WIDE BAY DISTRICT: Panorama Drive via Duke Rd, km SE of Cooroy, 26°27′S, 152°59′E, *Bean 5933* (BRI); Webster Rd, Goomboorian, 26°03′S, 152°45′E, *Bean 6501* (BRI); Old Ceylon Rd, 6 km SW of Cooroy, 26°27′S, 152°52′E, *Bean 7133* (BRI); South Amamoor Ck headwaters, Amamoor Ra., property of J. Williams adjacent to SF 435, 26°23′SD, 152°31′E, *Forster PIF15190* (BRI); Kin Kin, Eastern Branch Rd, property of A. & G. Black, *Forster PIF16124* (BRI); Coongara LA near Coongara Falls, 25°39′S, 152°00′E, *Grimshaw PG2206 & Turpin* (BRI); Booloumba Ck, Conondale SF, Mt Langeley, 26°40′S, 152°36′E, *Halford Q2170 & Hohnen* (BRI); Eastern Branch rd, Kin Kin, at head of Kin Kin Ck, 26°14′S, 152°55′E, Feb 1986, *Leiper s.n.* (BRI Accession [not AQ] 381026); Kin Kin (E Branch Rd), Oct 1986, *Leiper s.n.* (BRI AQ431639); Kin Kin, 26°16′S, 152°53′E, *Priest 3* (BRI); Kin Kin, on flats of creek, 26°16′S, 152°53′E, Oct 1989, *Priest s.n.* (BRI AQ457915); Sister Tree Ck, Kin Kin, 26°16′S, 152°53′E, Jan 1990, *Priest s.n.* (BRI AQ459431); Old Cooroy Rd, ca. 3 km NE of Eumundi, 26°27′S, 152°57′E, *Sharpe 4918 & Thomas* (BRI, MEL); Bingera, 24°58′S, 152°12′E, *Smith 4139* (BRI, CANB); Lenthalls Dam, Wongi SF, ca. 22 km NW of Maryborough, 25°24′S, 152°31′E, *Sparshott KMS888 & Turpin* (BRI); Top of Woowonga Ra., ca. 15 km NE of Biggenden, 25°27′S, 152°08′E, *Young 229 & Randall* (BRI).—MORETON DISTRICT: Enoggera Ck, Oct 1875, *Bailey s.n.* (BRI AQ93438); Pine R., *Bailey s.n.* (BRI 93327); Browning Property, near Cogill Rd, Buderim, 26°41′S, 153°04′E, *Bean 8867* (BRI); Moggill SF, 27°S, 152°E, Mar 1980, *Bird s.n.* (BRI AQ331061);Woogaroo Ck, Goodna, Ipswich, 27°S, 152°E, Oct 1983, *Bird & Smith s.n.* (BRI AQ431244); Woogaroo Ck, Goodna, Ipswich, 27°S, 152°E, Feb 1986, *Bird s.n.* (CONN, BRI 431241, L, LE, NSW, SP, US); at end of Grand View Rd, Moggill SF, along Pullen Ck area, 27°S, 152°E, Jan 1988, *Bird s.n.* (BRI AQ437283, CANB); Pine River, *Fitzalan s.n.* (NSW 144689); Moggill SF, Pullen Ck, 27°31′S, 152°51′E, *Forster PIF14609 & Bird, PIF14610 & Bird* (BRI); Perry Rd, Image Flat, just N of

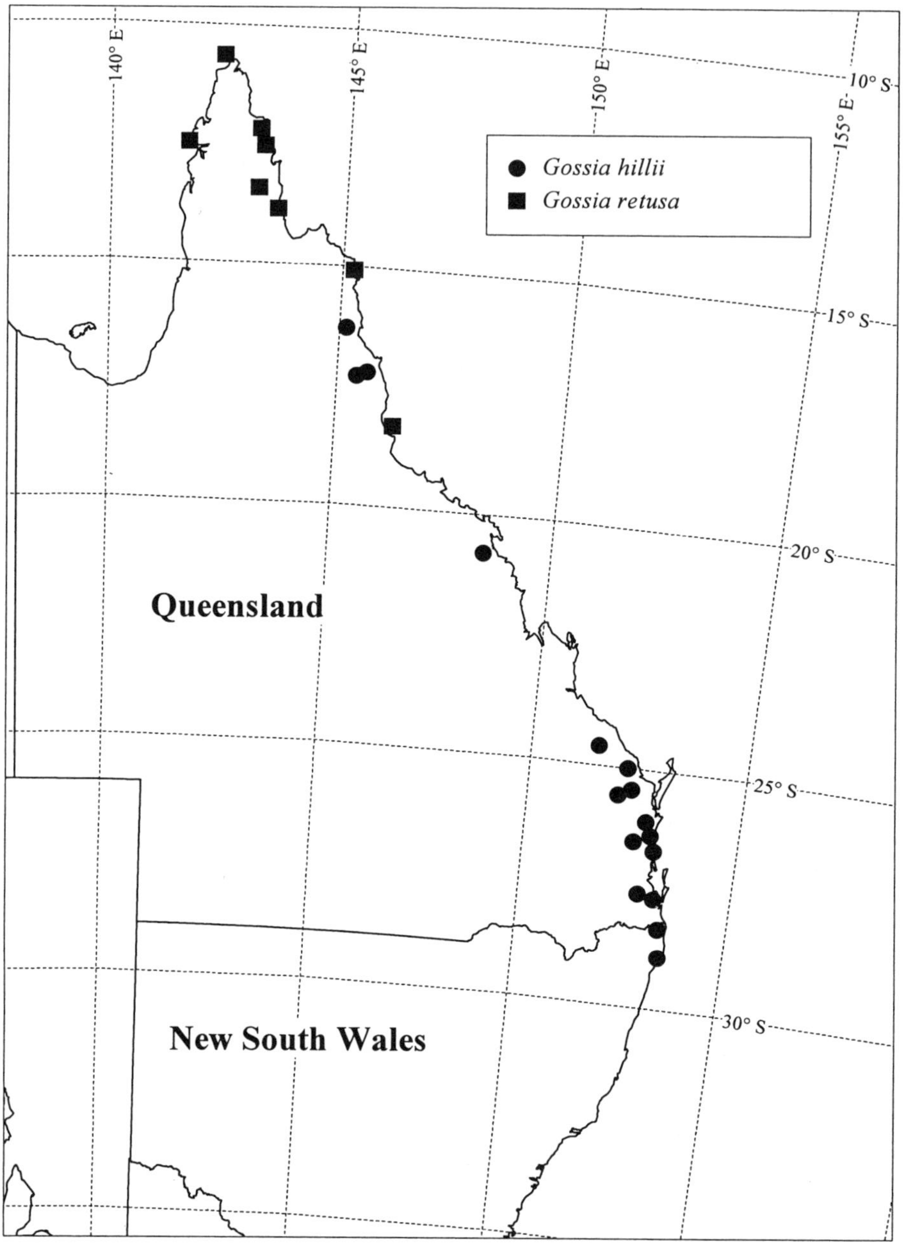

FIG. 14. Distribution of *Gossia hillii* and *G. retusa*.

Nambour, 26°36′11″S, 152°55′54″E, *Grimshaw G206 & Turpin* (BRI); Woogaroo Ck, E of Redbank Plains, 27°40′S, 152°53′E, *Guymer 2126* (BRI); S Main Camp rd, N of Palmwoods, 26°38′S, 152°58′E, *Guymer 2133* (BRI); Loganholme, S of Brisbane, 27°41′S, 153°11′E, Feb 1988, *Leiper s.n.* (BRI AQ437282); northern side of Tallebudgera Mt, *McDonald 2813 & Whiteman* (BRI); Kenmore area, Mar 1962, *Pearson s.n.* (BRI AQ093434); Sunday Ck, 28°S, 153°E, *Smith s.n.* (BRI AQ375338); 609 Grandview Rd, property of G. Nosworthy, Pullenvale, ca. 3.5 km from turnoff on Moggill Rd, 27°32′S, 152°50′E, *Snow 7747* (BRI, GREE); Moggill SF, NE corner, 0.1 km along first turnoff in Forest from Mill Rd, 0.8 km from intersection with Grandview Rd, 27°32′S, 152°50′E, *Snow & Snow 7742* (BRI, GREE); Ingleside State School, Tallebudgera Valley Rd, site 42, 28°10′S, 153°22′E, *Thomas* 78 *& Barry* (BRI); Murray's Rd, across creek from Logan Golf Course, Logan Shire, 27°39′S, 153°09′E, *Thomas 126 & Hoy* (BRI); 300 meters E of Daisy Hill, N of Chatswood Rd, 27°38′S, 153°10′E, *Thompson 126* (BRI); Goodna Scrub, Feb 1925, *White s.n.* (BRI-93439);Woogaroo Ck, S of Goodna, 27°40′S, 152°45′E, *Williams 83050* (BRI, CANB, NSW, RSA).—NEW SOUTH WALES: 8 km SE of Murwillumbah, 28°54′S, 153°26′E, *Guymer 1791 & Jessup* (BRI); Byrrill Ck Rd, S of Mt Warning, *Jessup 190* (BRI); Nimbin, Oct 1980, *Walker s.n.* (BRI AQ345399); 2 km E of Uki, Tweed R., 28°25′S, 153°19′E, Nov 1976, *Williams s.n.* (BRI AQ517575).

CULTIVATED: at Lot 29, Belah St, Mt Crosby, 27°31′S, 152°49′E, *Grimshaw G249* (BRI).

The combination of adaxially bright glossy leaves with a prominently raised midrib, inconspicuous glands on the abaxial leaf surface, and hairy cupulate flowers with sericeous linear bracteoles easily distinguishes *G. hilli* from other species of *Gossia*. The often longish pedicels are partially diagnostic as well. Some specimens (e.g., *Smith 4139*, BRI) have densely hairy petals.

The specific epithet honors Walter Hill (1820–1904), the first Superintendent of the Brisbane Botanic Gardens, who made the first collections in Queensland.

12. Gossia lewisensis N. Snow & Guymer, sp. nov.—TYPE: AUSTRALIA. Queensland: Cook District, Mount Lewis Road near Half Ton Creek crossing, about 30 km from Rex Hwy, 16°30′S, 145°16′E, 26 Nov 1988, *L. W. Jessup GJM20 & G. P. Guymer, W. J. McDonald* (holotype: BRI!; isotypes: AD! BRI! CANB! DNA! GREE! K! L! MEL! MO! NSW!).

Ramuli rotundati usque complanati; petioli atrobrunnei, folia juventute stipulis brevibus vaginantibus; folia pagina inferiore nervo medio pro parte maxima rufescenti. Flores pentameri.

Trees, 8–30 m tall. Bark of trunk flaky, sometimes becoming fluted or buttressed, brown to gray or sometimes cream-colored. Branchlets rounded to laterally compressed, not winged, brown to dark brown, smooth or soon becoming somewhat flaky, glabrous, oil glands few to common, indistinct. Leaves discolorous, slightly glossy or matte above, matte below. Stipules type C (see Morphology), but buds and adjacent interpetiolar regions of younger branches often enveloped by a thin, often obscure, ochreate stipule (Fig. 15C; see also Fig. 129b in *Flora of Australia*, 1999). Petioles 3–5 mm long, light brown to reddish brown, smooth above to slightly channeled, eglandular. Leaf blades 3.0–7.1 cm long, 1.2–3.6 cm wide, narrowly elliptic to ovate, base cuneate, apex acuminate, the tip pointed and often falcate, margins flat; adaxial surface glabrous, oil glands abundant and prominent (viewed with magnification); midvein flush; abaxial surface glabrous, oil glands common, prominent (viewed with magnification); secondary and intramarginal veins indistinct to prominent, the inner intramarginal vein 1–1.5 mm from edge. Inflorescence terminal or axillary, solitary, paired, or sometimes superimposed in fascicles; flower solitary or in short racemes or panicles; anthopodia absent, metaxyphylls present or absent. Peduncles (4–) 9–16 mm long, rigid, glabrous. Bracteoles 2 (but these sometimes only poorly differentiated from smaller, lateral, bractlike structures, and sometimes appearing somewhat fused to them), 1–1.5 mm long, 0.5–0.7 mm wide, narrowly triangular, rigid, scalelike, midrib present or absent, with tips shorter than base of calyx lobes, glabrous to minutely and sparsely sericeous along edges, persistent or caducous in fruit. Hypanthium 2–3 mm long, obconic to cupulate, glabrous or occasionally sparsely sericeous, oil glands abundant. Calyx lobes 5, 1–3 mm long, broadly ovate to rounded, apex obtuse, greenish; adaxial surface sericeous to villous; abaxial surface glabrous. Petals 5, 2.5–5.2 mm long, 3.5–5.5 mm wide, widely ovate to transversely elliptic (the edges repand), white to yellowish white; adaxial surface sparsely sericeous to villous (hairs somewhat ferrugineous); abaxial surface sparsely villous (hairs ferrugineous), oil glands few to common (but faint). Stamens 60–120, included, filaments 3–6 mm long; staminal disk short-villous (hairs somewhat ferrugineous); anthers 0.5–1 mm long, globose to subcylindrical, the tip of connectives eglandular or with a single gland (and sacs sometimes with minute glands). Style 2.5–5.5 mm long, glabrous to sparsely sericeous (and glandular below); ovary apex somewhat hairy; ovules 10–12 per placenta, placed irregularly. Berry 6.5–9.5 mm long, 7.5–10.5 mm in diameter, subglobose, base rounded, glabrous, dark red to blue-black. Seeds in mature fruit 1–8; subglobose to reniform, smooth, ivory-colored, micropylar operculum absent, seed coat hard, bony. Embryo circinate, starchy, oil glands absent or present but faint, endosperm present; hypocotyl relatively thick, tip near radicle swollen, tip at same plane as embryo; cotyledons straight, shorter than hypocotyl, folded back towards hypocotyl. Fig. 15.

Phenology. Flowering from October through December; fruiting November through August.

Distribution (Fig. 16). Northeastern Queensland; occasional in simple notophyll vine forests and rainforests on granite; 880–1200 m.

Conservation status. The species is considered Vulnerable (criterion C2a, Species Survival Commission 1994).

ADDITIONAL SPECIMENS EXAMINED. **Australia**. QUEENSLAND. COOK DISTRICT: Pinnacle Track, 3.5 km W of Karnak, 16°23′S, 145°18′E, *Forster PIF15536 et al.*, *PIF15543 et al.* (BRI); 10 km along Mt Lewis Rd, TR 66, Bushy LA, 16°35′S, 145°17′E, *Forster PIF15613 et al.* (BRI, QRS); Mt Fraser area, Eyres LA, 16°36′S, 145°16′E, *Forster PIF15615 et al.* (BRI); SFR143, North Mary LA, 16°32′S, 145°15′E, *Gray 831* (BRI, CANB); SFR143, North Mary LA, 16°32′S, 145°15′E, *Gray 1503* (BRI, CANB); SFR143 South Mary LA, 16°33′S, 145°17′E, *Gray 1558* (BRI, CANB, QRS); SFR143, North Mary LA, 16°32′S, 145°16′E, *Gray 2844* (BRI, QRS); Zarda LA, near Zarda Clearing, 16°27′E, 145°15′E, *Hyland 2893* (BRI, QRS); SFR143, Windmill LA, 16°34′S, 145°17′E, *Hyland 11693* (BRI, QRS); SFR 144, Whypalla, Bower Bird LA, 16°15′S, 145°01′E, *Hyland 25385* (BRI); Mt Misery, E of Mt Spurgeon, 15.7 km NE of Mt Carbine, 16°26′S, 145°13′S, *Jessup et al. GJM754* (BRI); 29 km along Mt Lewis Rd from Mossman–Mt Molloy Rd turnoff, 16°31′S, 145°16′E, *Jessup et al. GJM5168* (BRI); SFR 144, Windsor Tableland, 16°15′S, 145°02′E, *Moriarty 2571* (BRI, CANB); E/P 18 North Mary LA, R143 Mt Lewis, 16°30′S, 145°16′E, *Sanderson 460* (QRS); SFR 144, Exp. E/P 30, 16°17′S, 145°05′E, *Sanderson 1036, 1086* (QRS); SFR 144 Fantail LA, EP 44, 16°12′S, 145°05′E, *Sanderson 1792* (QRS); SFR 144, Windsor Tableland, 16°15′S, 145°04′E, *Stocker 1618* (BRI, CANB, QRS); Mt Lewis near Julatten, 16°35′S, 145°17′E, *Tucker 45* (BRI).

The ochreate stipules characteristic of the young leaves can be somewhat cryptic, but help to distinguish *G. lewisensis* from its congeners. In addition, the branchlets are round to flattened, the midvein of the abaxial leaf surface is mostly dark reddish brown, the petal hairs are ferrugineous, and the anther sacs often have minute oil glands. The bark is said to be very thin and layered when dead (Hyland et al. 1999).

The specific epithet refers to the type locality on Mt Lewis in northeastern Queensland.

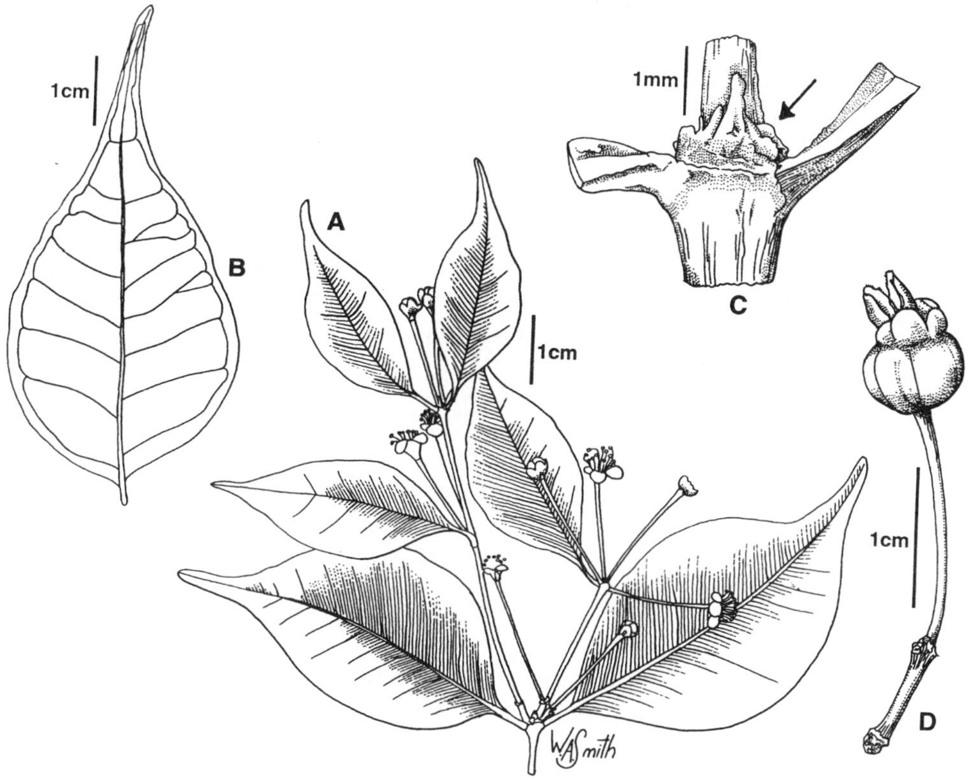

FIG. 15. *Gossia lewisensis.* A. Flowering branch. B. Leaf. C. Node; the arrow points to the ochreate (=sheathing) stipules of youngest leaves. D. Fruit on peduncle, with remnant petals at apex. (Based on: A, *Dansie 20172*; B, C, *Tucker 45*; D, *Hyland 25385*.)

13. Gossia punctata N. Snow & Guymer, sp. nov.—TYPE: AUSTRALIA. Queensland: Moreton District, Kamarun Lookout on road to O'Reillys in Lamington National Park, edge of picnic area, 28°10′S, 153°06′E, 29 Nov 1997, *N. Snow & J. Bradford 7471* (holotype: BRI!; isotypes: AD! B! BRI! CANB! DNA! GREE! HO! K! L! MEL! MO! NSW! PERTH! SING! UPS! US!). [All duplicates from a single individual.]

Ramuli rotundati; ad instar *G. hillii* autem folia pagina inferiore glandibus oleosis conspicuis dense obtecta. Flores pentameri hypanthio basi certe decrescenti; petalis pubes centibus non nisi margine.

Large shrubs to trees, 3–15 m tall. Bark of trunk fissured to tessellated and sometimes flaky, gray to dark brown or reddish. Branchlets rounded to laterally compressed, not winged, light brown to light gray, bark smooth, sparsely sericeous but becoming glabrous, oil glands common, indistinct. Leaves concolorous or somewhat discolorous, glossy above, matte below. Stipules type A or type C (see Morphology). Petioles 1–3 mm long, light to dark brown, smooth above to slightly channeled, eglandular. Leaf blades 1.3–4.5 cm long, 0.8–2.0 cm wide, narrowly elliptic to obovate, base cuneate, apex acute to acuminate, the tip rounded or rarely falcate, margins flat; adaxial surface glabrous, oil

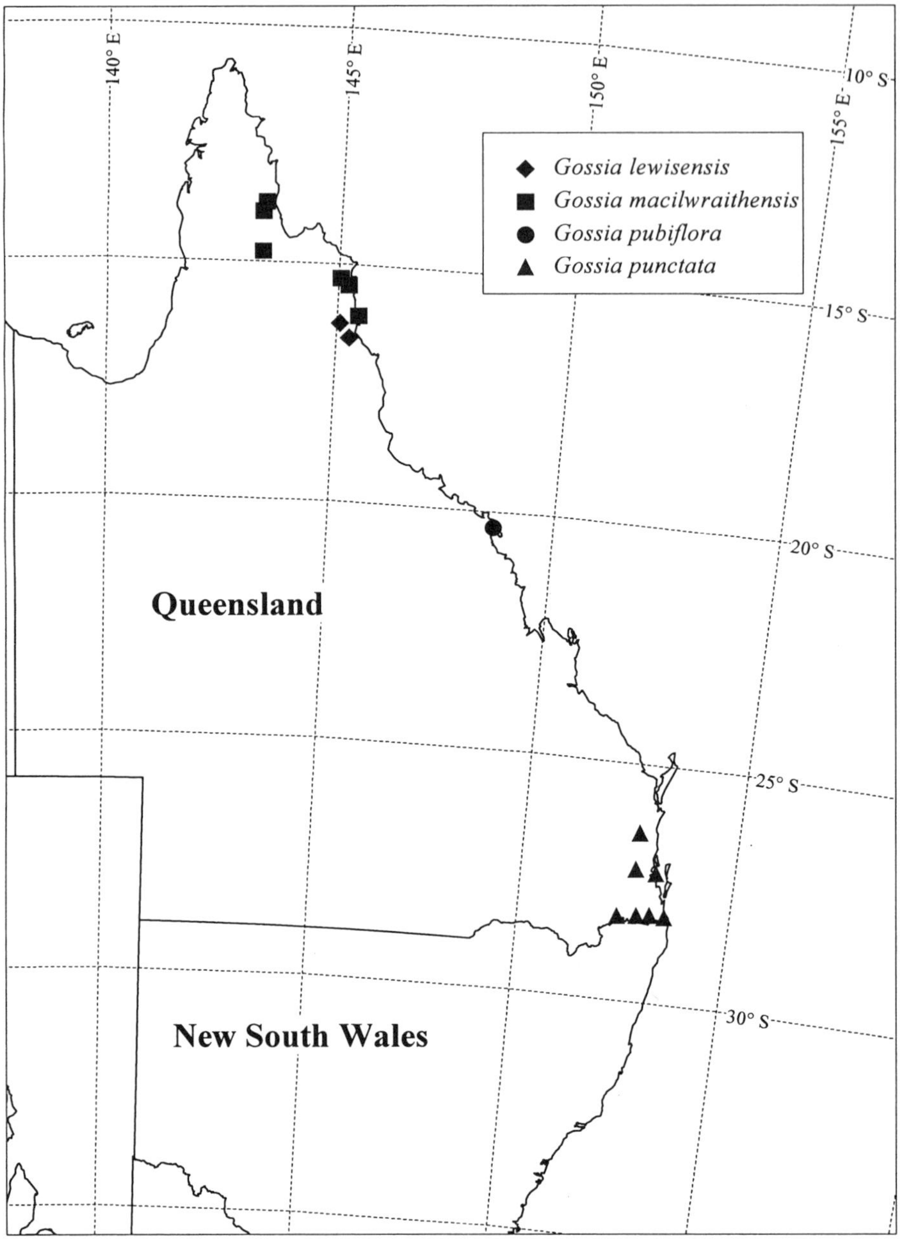

FIG. 16. Distribution of *Gossia lewisensis, G. macilwraithensis, G. pubiflora*, and *G. punctata*.

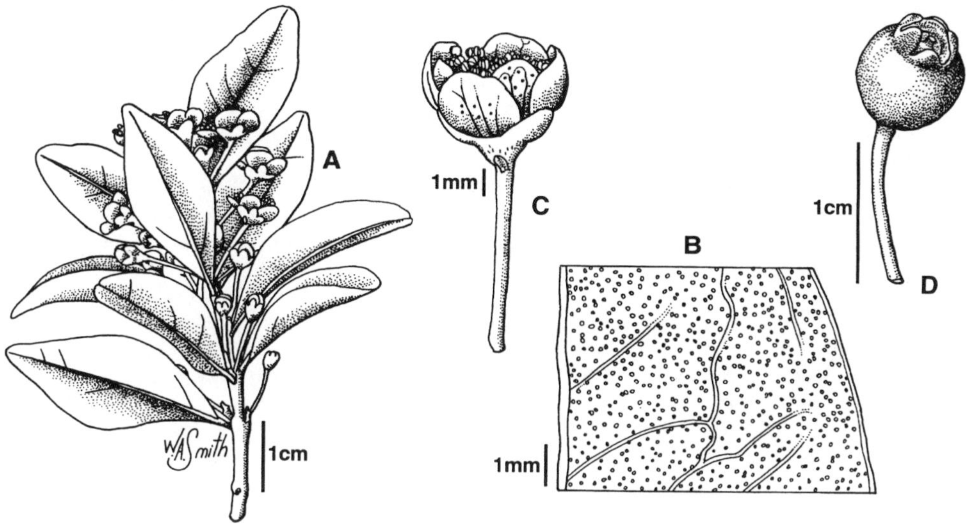

FIG. 17. *Gossia punctata*. A. Flowering branch. B. Detail of abaxial leaf surface. C. Flower. D. Fruit and portion of peduncle. (Based on: A–C, *Jessup 155*; D, *Snow 7471 & Bradford*.)

glands not visible to the naked eye; midvein flush; abaxial surface glabrous, oil glands abundant, prominent; secondary and intramarginal veins not visible to the naked eye to indistinct, the intramarginal vein ca. 1 mm from edge. Inflorescence terminal or axillary, solitary; flowers solitary or in racemes or panicles; anthopodia and metaxyphylls present. Peduncles 10–17 mm long, rigid, glabrous to sparsely sericeous. Bracteoles 2 (occasionally disjunct opposite), 0.9–1.1 mm long, ca. 0.5 wide, linear, rigid, scalelike without midrib, with tips shorter than base of calyx lobes, sparsely sericeous, caducous in fruit. Hypanthium 1.5–2 mm long, obconic, sparsely to densely sericeous, oil glands common. Calyx lobes 5, 1.5–3 mm long, broadly ovate to rounded, apex obtuse, greenish; adaxial surface glabrous to sparsely sericeous and ciliate on margins; abaxial surface glabrous to sparsely sericeous. Petals 5, 3.5–5.5 mm long, 3.5–4 mm wide, widely ovate to widely obovate or oblate, yellowish white; adaxial surface glabrescent to sparsely sericeous; abaxial surface glabrous (edges slightly ciliate), oil glands few and faint. Stamens 75–145, included, filaments 4–5 mm long; staminal disk glabrous; anthers 0.5–1 mm long, subglobose to globose, the tip of connective with a single gland. Style 3–4 mm long, glabrous to sparsely sericeous; ovary apex sparsely hairy; ovules 5–15 per placenta, placed irregularly. Berry 7–12 mm long, 5–12 mm in diameter, subglobose to subcylindrical, base rounded, glabrous, dark purple to black. Seeds in mature fruit 1–4; somewhat rounded and compressed, smooth, ivory to light brown, micropylar operculum present, seed coat hard, bony. Embryo circinate, starchy, oil glands absent, starchy, endosperm absent; hypocotyl relatively thick, tip near radicle swollen, tip at same plane as embryo; cotyledons circinate, shorter than hypocotyl, pointing forward or somewhat reflexed. Fig. 17.

Phenology. Flowering from October through January; fruiting January through April.

Distribution (Fig. 16). From northern New South Wales to southeastern Queensland; locally common in subtropical rainforest to microphyll vine forest or woodlands of *Eucalyptus propinqua* H. Dean & Maiden, creek banks to ridgetops; 140–400 m.

Conservation status. The species should be considered Vulnerable (criteria C2a,

Species Survival Commission 1994) until collections from several additional sites suggest otherwise.

ADDITIONAL SPECIMENS EXAMINED. **Australia.** QUEENSLAND. WIDE BAY DISTRICT: Imbil, 26°27′S, 152°40′E, *Guymer 2090* (BRI); Kin Kin, Jan 1916, *White s.n.* (BRI AQ93437, BRI AQ93432).—MORETON DISTRICT: Camp Ck, 12 km SE of Rathdowney, 28°S, 152°E, Mar 1986, *Bird s.n.* (BRI AQ431238); Camp Ck–Glenapp District, 15 km SE of Rathdowney, 28°18′S, 152°52′E, Mar 1986, *Bird s.n.* (BRI, CANB, L, K, MEL, MO, NSW); Lacey's Ck track, D'Aguilar Ra. SF, 27°13′11″S, 152°41′58″E, *Grimshaw G255* (BRI); Moreton Bay, 1862, *Hill s.n.* (MEL 726445); Upper Brookfield, Brisbane, 27°S, 152°E, *Jessup 155* (BISH, BRI, CANB, MEL, MO, NSW), *178* (BRI); Kamarun Lookout, on rd to O'Reillys, 28°1′S, 153°E, *Leiper s.n.* (BRI AQ459195); Shaw's Pocket, 7 km S of Beenleigh, 27°15′S, 153°10′E, *Leiper 2 & Thompson* (BRI); roadside ca. 2 km S of Kamarun Lookout, Lamington NP Rd, 28°10′S, 153°06′E, *McDonald 6174* (BRI, MEL); ridge SE of Mt Sim Jue, 27°15′S, 152°40′E, *Phillips 202* (BRI); Eumundi, Nov 1894, *Simmonds s.n.* (BRI AQ93328); end of Brock Rd, near Nicoll Scrub NP, 28°12′S, 153°24′E, *Snow & Bradford 7459* (BRI, CANB, GREE, L, K, MEXU, MO, NSW, US).—DARLING DOWNS DISTRICT: The Head, lower slopes of Wilsons Peak on rabbit fence, 28°14′S, 152°28′E, *Forster PIF15692 & Leiper* (BRI).—NEW SOUTH WALES. South branch of Hopping Dicks Ck, NW of Limpinwood, 28°18′S, 153°11′E, *Guymer 1806 & Jessup* (BRI, CANB); Cox's Rd, *Hayes 4* (BRI); Unumgar SF, *Jones s.n.* (BRI Q093425); Toonumbar SF, 25 km WNW of Kyogle, *Webb et al. 2163* (BRI).

The abaxial leaf surface of *G. punctata* is densely covered with oil glands. The adaxial leaf surface is glossy, like that of *G. hillii*, but the midvein is flush.

The specific epithet refers to the prominent punctate oil glands on the leaves.

14. Gossia dallachiana (F. Mueller ex Bentham) N. Snow & Guymer, comb. nov. *Eugenia dallachiana* F. Mueller ex Bentham, Fl. austral. 3: 287. 1867. *Austromyrtus dallachiana* (F. Mueller ex Bentham) L. S. Smith, Proc. Roy. Soc. Queensland 67: 35. 1956.—TYPE: AUSTRALIA. Queensland: North Kennedy District, Rockingham Bay, *J. Dallachy s.n.* (holotype: K!, photo: BRI!; isotypes: BRI! MEL-2 sheets!).

Shrubs or trees, 3–8 m tall. Bark of trunk smooth to flaky, light brown to light gray, sometimes mottled with pink or brown (fide Hyland & Whiffin, 1993, and Hyland et al., 1999). Branchlets rounded, not winged, whitish to light brown or light gray, smooth to slightly flaky, glabrous, oil glands common, not visible to the naked eye to indistinct. Leaves discolorous or becoming concolorous, matte above and below. Stipules type B (see Morphology), sometimes indistinct or absent. Petioles 4–10 mm long, light brown to black, smooth above or slightly channeled, eglandular. Leaf blades 8.5–20.0 cm long, 4.5–12.0 cm wide, elliptic to broadly elliptic, base cuneate or occasionally rounded, apex acute, the tip rounded, margins flat; adaxial surface glabrous, oil glands common to abundant, prominent (viewed with magnification); midvein mostly flush or slightly raised; abaxial surface glabrous, oil glands common to abundant, prominent (viewed with magnification); secondary veins of lower surface prominent; intramarginal veins of lower surface prominent, innermost intramarginal veins on abaxial leaf surface 4–14 mm from margin. Inflorescence terminal or axillary, solitary; flowers in panicles; anthopodia and metaxyphylls absent. Peduncle 9–14 mm long, rigid, glabrous. Bracteoles 2, ca.1.5 mm long, 0.5–0.8 mm wide, linear to narrowly elliptic, rigid, scalelike or slightly foliaceous, midrib mostly absent or rarely present, the tips shorter than base of calyx lobes, glabrous or minutely and sparsely sericeous along edges, caducous in fruit. Hypanthium 3–5 mm long, obconic, glabrous, oil glands abundant. Calyx lobes 4, 1.5–3.5 mm long, broadly ovate, apex obtuse, green or somewhat petaloid; adaxial surface glabrous to densely sericeous; abaxial surface glabrous.

Petals 4, 5.5–10 mm long, 5–11.5 mm wide, widely ovate to widely obovate or circular, white; adaxial surface sparsely sericeous to villous; abaxial surface sparsely villous (particularly near margins), oil glands common but faint. Stamens (60–) 150–200, included, filaments 4–7 mm long; staminal disk glabrous; anthers 0.5–0.9 mm long, globose to subcylindrical, the tip of connective with a single gland (sometimes faint). Style 5–8.5 mm long, glabrous; ovary apex glabrous; ovules 1–6 per placenta, placed irregularly. Berry 8–16 mm long, 8–16.5 mm in diameter, subglobose to subcylindrical, base rounded or occasionally tapered, glabrous, dark blue to black. Seeds in mature fruit 1, more or less round, smooth to strongly irregular with indentations, ivory-colored, micropylar operculum absent, seed coat hard and bony. Embryo circinate, starchy, oil glands present, endosperm present; hypocotyl relatively thick, tip near radicle swollen, at same plane as embryo; cotyledons mostly straight, shorter than hypocotyl, pointing forward.

Phenology. Flowering from January through May, peaking in March; fruiting March through October.

Distribution (Fig. 18). Central and northeastern Queensland; occasional to locally common in understory of rainforest; sea level to 1100 m.

Vernacular name. Lignum.

ADDITIONAL SPECIMENS EXAMINED. **Australia**. QUEENSLAND. COOK DISTRICT: Towalla Rd, 17°27′S, 145°43′E, *Cooper & Cooper 1120* (QRS); Tinaroo (Lamb) Ra., ca. 8 km NNE of Tinaroo Dam, 17°07′S, 145°35′E, *Crome 483* (CANB); Forestry Reserve 310, Compt. 21, Atherton district, *Dansie 2013* (BRI); head of Peach Ck, Leo Ck rd, McIlwraith Ra., 13°44′S, 143°20′E, *Fell 2282* (BRI); Middle Claudie R., 12°44′S, 143°14′E, *Forster PIF13567 et al.* (BRI); Westcott Rd, Topaz, 17°24′S, 145°42′E, *Forster PIF13771* (BRI); SFR 755, Barong LA, 17°31′S, 145°50′E, *Gray 352* (BRI, CANB, MEL, MO, QRS); SFR 310, Parish of Gadgarra, Tardent LA, 17°18′S, 145°43′E, *Gray 4706* (QRS); ca. 3.5 km along Boonjee Rd, 17°22′S, 145°44′E, *Guymer 2062* (BRI); 2.5 km along Boonjee Rd, E of Lamins Hill, 17°22′S, 145°44′E, *Guymer 2065* (BISH, BRI, MO); SFR 143 North Mary LA, 16°30′S, 145°15′E, *Hyland 3424RFK* (BRI, CANB); SFR 194, 17°15′S, 145°25′E, *Hyland 4988* (BRI, QRS); SFR 756, East Downey LA, 17°40′S, 145°50′E, *Hyland 5607* (BRI, QRS); Lankelly Ck, 13°55′S, 143°20′S, *Hyland 5365* (BRI); SFR 310, Gadgarra, Windin LA, 17°23′S, 145°44′E, *Hyland 14095, 14236* (QRS); SFR 607, 17°05′S, 145°35′E, *Irvine 761* (BRI, CANB, NSW, QRS); Mt Baldy 5.4 km W of Atherton near Walsh Falls, site 36, 17°17′S, 145°25′E, *Jessup GJM1975 et al.* (BRI); FR 756, Campbell LA, 17°40′S, 145°35′E, *Martin 3002 & Bardsley* (BRI); SFR 185, Robson LA, 17°10′S, 145°37′E, *Risley 367* (QRS); Woopen Ck SF, 17°30′S, 145°50′E, *Sankowsky 552 & Sankowsky* (BRI); Windsor Tableland, 16°15′S, 145°20′E, *Sankowsky 583 & Sankowsky* (BRI); Leo Ck Rd (TR 14), 13°45′S, 143°20′E, *Stocker 950* (BRI); Palmerston NP, *Thorne 27641* (QRS, RSA), Nesbit R., 13°26′S, 143°10′E, *Tracey 14230* (BRI); Bartle Frere Rd, ca. 1 km past Swipers Flat Rd, Boonjie turnoff, 17°S, 145°E, *Tucker MCT95* (BRI); Lower Palmerston Hwy, W of Innisfail, 17°34′S, 145°45′E, *Webb & Tracey 11448* (BISH, BRI, CANB, MO); Boonjie, W of Mt Bartle Frere, *Webb & Tracey 6578* (BRI); McIvor R., N of Cooktown, *Webb & Tracey 7803* (BRI); headwaters of Massey Ck near old mining site, McIlwraith Ra., ca. 13°50′S, 143°20′E, *Webb & Tracey 9236* (BRI); headwaters of Massey Ck near old mining site, McIlwraith Ra., ca. 13°50′S, 143°20′E, *Webb & Tracey 9236* (BRI); headwaters of Lankelly Ck on western fall of McIlwraith Ra., ca. 13°52′S, 143°20′E, *Webb & Tracey 9575* (BRI); Gadgarra, *White 1574* (BRI); SFR 310, Swipers LA, 17°22′S, 145°47′E, *Whiffin 631 & Risley* (QRS); Malanda, Jan 1923, *White s.n.* (BRI AQ093386).—NORTH KENNEDY DISTRICT: Yuccabine Ck, SF 344 Kirrama LA, 28 km NW of Cardwell, 18°12′S, 145°45′E, *Jessup et al. GJM2373* (BRI); Culpa LA, upper catchment of Culpa Ck, 13.5 km SSE of Koombooloomba township, 17°56′S, 145°36′E, *Jessup et al. GJM5215* (BRI); Rockinghams Bay, *Mueller s.n.* (MEL 67167, 67168, 67169); between Society Flat and Yuccabine Ck, *Smith 3184* (BRI, CANB, MO); Tully Falls, 17°47′S, 145°33′E, *Smith 4742* (BRI); Tableland LA, SFR 251, 17°45′S, 145°33′E, *Stocker 905* (BRI, QRS); Cannabullen via Ravenshoe, *Webb & Tracey 7634* (BRI).—SOUTH KENNEDY DISTRICT: Calen near Mackay, *Webb & Tracey 7803* (BRI).

Gossia dallachiana is easily distinguished from other species in the genus by its large leaves, which have the inner intramarginal vein placed up to 15 mm from the margin. The petals can be up to 10 mm long.

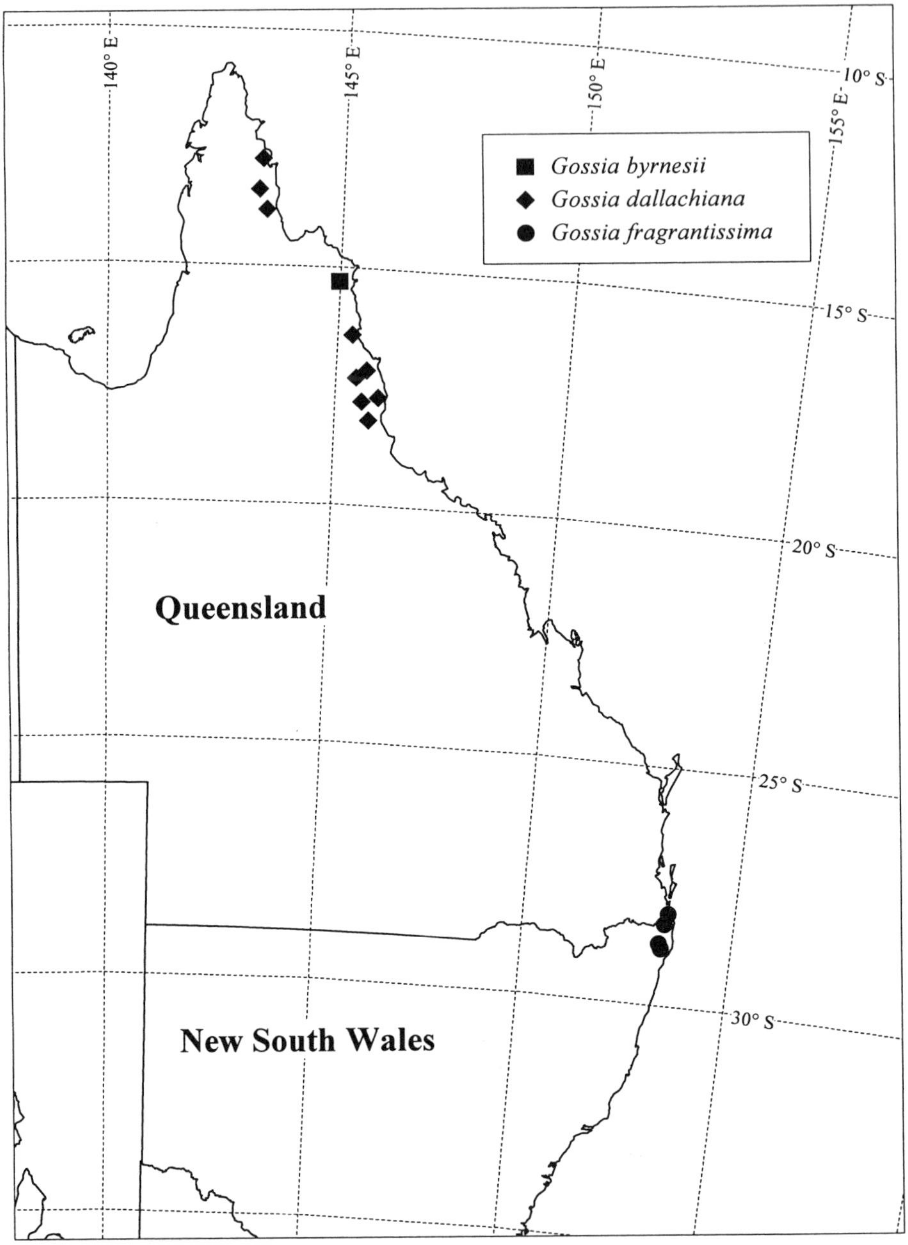

FIG. 18. Distribution of *Gossia byrnesii, G. dallachiana*, and *G. fragrantissima*.

Hyland et al. (1999) report that *G. dallachiana* seldom exceeds 30 cm dbh and occurs in the understory. The occasional presence of asteroid clusters of bracts in or near the leaf axils is evidently due to disease (e.g., *Forster PIF13771*, BRI).

The specific epithet honors John Dallachy (ca. 1808–1871), who preceded Ferdinand von Mueller as Superintendent of the Melbourne Botanic Gardens, and who collected in the area of the current city of Rockhampton, Queensland, in 1863.

15. Gossia grayi N. Snow & Guymer, sp. nov.—TYPE: AUSTRALIA. Queensland: Cook District, State Forest Reserve 310, Gillies Logging Area, 17°22′S, 145°42′E, 11 Jul 1979, *B. Gray 1484* (holotype: BRI!; isotypes: BRI! CANB! QRS!, also spirit collections).

Ramuli rotundati; petioli vivi virides, in sicco nigrescentes; folia pagina superiore nitida, nervo medio elevato, apice retuso. Flores tetrameri.

Trees, 10–18 m tall. Bark of trunk smooth, somewhat flaky to slightly fissured, light brown or gray, sometimes mottled with orangish or pinkish. Branchlets rounded, not winged, light brown to reddish brown, smooth or flaky, glabrous, oil glands few to common, indistinct. Leaves slightly discolorous, glossy or matte above, matte below. Stipules absent. Petioles 4–6 mm long, green (fresh) to dark brown or mostly black (dry), channeled or smooth above, eglandular. Leaf blades 6.0–11.5 cm long, 2.8–5.5 cm wide, elliptic to broadly elliptic, base cuneate, apex acute, the tip rounded or rarely retuse, margins flat; adaxial surface glabrous, oil glands abundant, prominent (viewed with magnification); midvein raised (particularly proximally); abaxial surface glabrous, oil glands abundant, prominent (viewed with magnification); secondary veins prominent; intramarginal veins indistinct to prominent (inner intramarginal vein more prominent), the inner intramarginal vein 2.5–4 mm from edge. Inflorescence terminal or axillary, solitary to fascicled; flowers solitary or in racemes; anthopodia and metaxyphylls absent. Peduncles 10–17 mm long, rigid, glabrous. Bracteoles usually absent or 2, 1–1.5 mm long, ca. 0.5 mm wide, narrowly triangular to triangular, scalelike, midrib absent or faint, with tips shorter than base of calyx lobes, glabrous, caducous in fruit. Hypanthium 1.5–2 mm long, obconic, glabrous, oil glands common. Calyx lobes 4, 1.5–2 mm long, triangular to broadly ovate, apex obtuse, petaloid; adaxial surface densely sericeous; abaxial surface glabrous. Petals 4, 4–6 mm long, 3–4 mm wide, widely ovate to obovate, white to yellowish white; adaxial surface glabrous but shortly ciliate on margins; abaxial surface glabrous, oil glands few. Stamens 50–60, included, filaments 2.5–5 mm long; staminal disk glabrous to glabrescent; anthers 0.6–0.8 mm long, subglobose to globose, the tip of connective eglandular. Style 4–6 mm long, glabrous; ovary apex glabrous; ovule number and placement on placenta unknown. Berry 7–18 mm long, 9–16 mm in diameter, subglobose to globose, base rounded, glabrous, dark blue-black. Seeds in mature fruit 4–8; irregularly reniform and wedge-shaped in profile, smooth, light brown, micropylar operculum present, seed coat leathery. Embryo circinate, starchy, oil glands absent, starchy, endosperm present, milky white and gelatinous; hypocotyl relatively thick, tip near radicle somewhat swollen, tip at same plane as embryo; cotyledons straight, shorter than hypocotyl, folded back towards hypocotyl or pointing forward. Fig. 19.

Phenology. Flowering in March; fruiting June through December.

Distribution (Fig. 9). Northeastern Queensland; rainforests and simple or mixed notophyll vine forests, sometimes in secondary forests, on granitic or metamorphic chlorite schists; 300–1200 m.

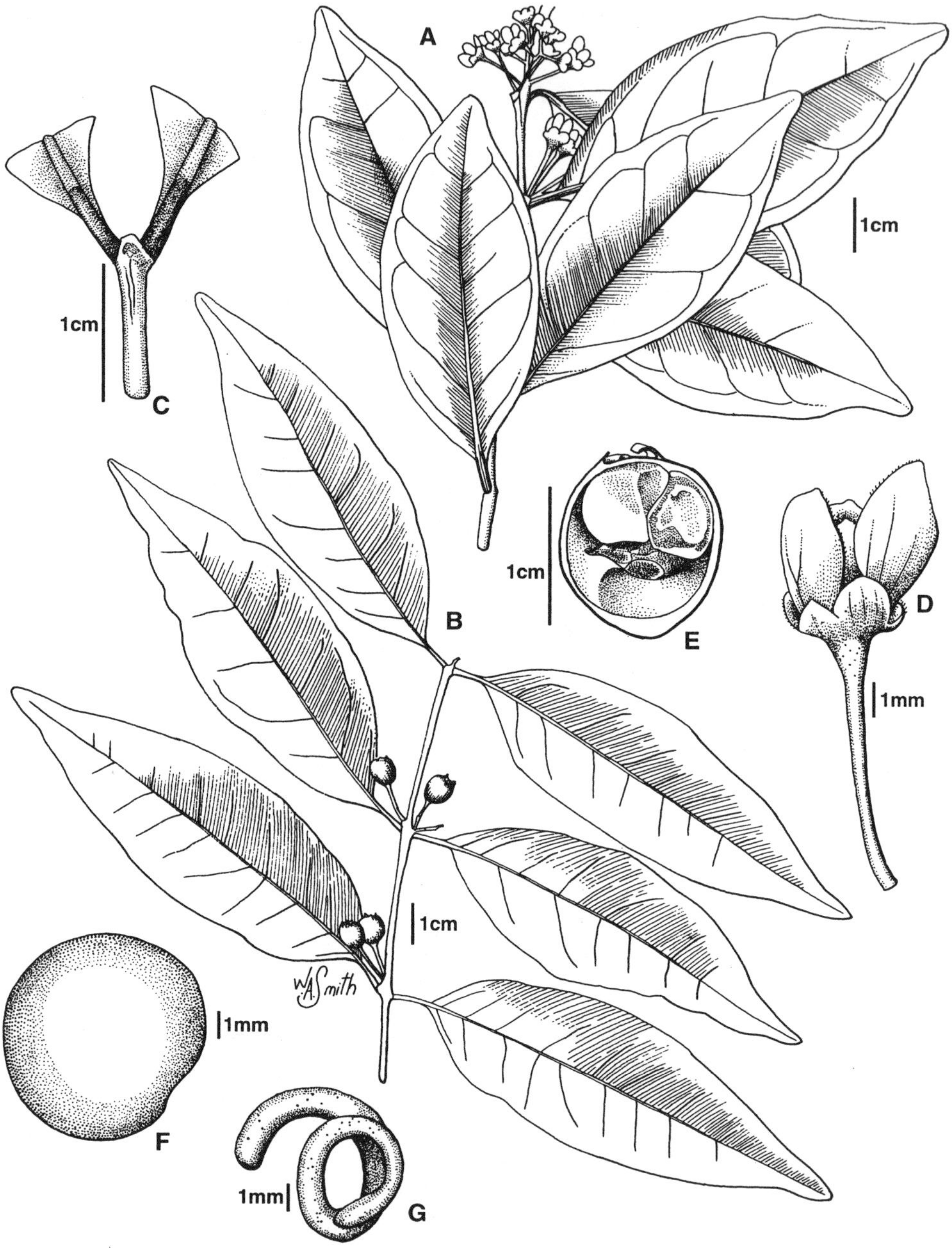

FIG. 19. *Gossia grayi.* A. Flowering branch. B. Fruiting branch. C. Dark petioles of opposite leaf pair. D. Flower in early anthesis. E. Fruit in longitudinal section, showing two seeds attached to centrally located placenta. F. Seed. G. Embryo. (Based on: A, C–D, *Gray 380*; B, *Gray 5244*; E–G, *Gray 1509*.)

Conservation status. *Gossia grayi* is probably not Vulnerable, but additional data is needed on its abundance and overall distribution before it status can be decided.

Vernacular name. Lignum.

ADDITIONAL SPECIMENS EXAMINED. **Australia.** QUEENSLAND. COOK DISTRICT: SFR 700, Gillies LA, 17°13′S, 145°41′E, *Gray 380* (BRI, CANB), *604* (BRI); *1462* (BRI), *1509* (BRI), *1510* (BRI, CANB), *1518* (BRI); Leo Ck Rd (TR 14), 13°45′S, 143°22′E, *Gray 5244* (QRS); SFR 41 (near Crater), 17°25′S, 145°25′E, *Hyland 535* (BRI); SFR 194, 17°20′S, 145°25′E, *Hyland* 1620 (BRI); Mt Carter, 13°00′S, 143°15′S, *Hyland 7536* (BRI); SFR 607, Emerald LA, 17°05′S, 145°35′E, *Irvine 537* (BRI, CANB, QRS [spirits]); 0.4 km before W Spencer Ck Forestry Camp, SF 144, Cockatoo LA, 0.3 km E Mt. Carbine, Mt Windsor Tableland, 16°15′S, 145°02′E, *Jessup GJM1286 et al.* (BRI); SFR 700, Gillies LA, 17°14′S, 145°41′E, *Moriarty 2084* (QRS); Emerald LA,17°09′S, 145°38′E, *Sanderson 320* (QRS); SFR 144, 16°17′S, 145°05′E, *Sanderson 1087* (QRS); McIlwraith Ra. (TR 14), 13°45′S, 143°20′E, *Unwin 370* (QRS); Forestry Reserve 99, Western, a few mi SW of Atherton, *Volck QF57/104* (BRI); Davies Ck, Lamb Ra. between Mareeba and Kuranda, *Webb & Tracey 6722* (BRI); Danbulla, *Webb & Tracey 7213* (BRI); Herberton Ra., *Webb & Tracey 7961* (BRI); SFR 310 Mt Lewis, 16°35′S, 145°16′E, *Webb & Tracey 10526* (BRI).—NORTH KENNEDY DISTRICT: SFR 65, 5.5 km along Cupla Rd from Koombooloomba Rd, site 46, 17°52′S, 145°36′E, *Ford 3227 & Cooper* (GREE); Kirrama Ra., W of Kennedy, *Smith 3213* (BRI).

Specimens of *G. grayi* are often difficult to determine with certainty. The bark varies from brown to gray with patches or hues of red or pink. The petioles are green when fresh but dry to nearly black (A. Ford, pers. comm.), and the dull, light green leaf blades have a cuneate base; stipules are absent. The hypanthium is obconic, and bracteoles are absent in most inflorescences. The flowers are 4-merous and have petaloid sepals.

Hyland et al. (1999) discerned two elements in the entity here described as *G. grayi.* One, "Leo Creek," has a more northerly range, from approximately Coen and Silver Plains northward. The other variant, "Gillies," occurs about the Windsor Tableland south to Malanda. We were unable to define the populations from these disjunct areas as two taxa separated by fixed morphological differences. According to A. Ford (pers. comm.), the bark of young specimens is smooth but dull colored, whereas with age it turns pinkish or at least has large areas of pink. Its apperance resembles species of "python bark" trees in *Gossia*. The reflectance of the upper leaf surface varies from slightly glossy to mostly densely matte.

The epithet honors Bruce Gray (b. 1939) of CSIRO in Atherton, Queensland, who has made many important collections of this new species and others in *Gossia*.

16. Gossia macilwraithensis N. Snow & Guymer, sp. nov.—TYPE: AUSTRALIA. Queensland: Cook District, Timber Reserve 14, Kesteven, 13°44′S, 143°20′E, 22 Oct 1981, *B. Hyland 11148* (holotype: BRI!; isotype: QRS!, also spirit collections).

Ramuli rotundati; folia pagina superiore implexa nervo medio manifeste elevato et in petiolum extenso, glandibus oleosis non visibilibus, apice obtuso usque late acuto. Flores tetrameri.

Shrubs to trees, 3–15 m tall. Bark of trunk smooth to rough and sometimes fluted, whitish to gray or brown, sometimes orangish. Branchlets rounded, not winged or rarely winged (but the wings thin and soon peeling away), light brown to gray or orangish, smooth to flaky or stringy, glabrous to sparsely sericeous, oil glands few, not visible with the naked eye or indistinct. Leaves concolorous, matte above and below. Stipules type A (see Morphology), often extending laterally between petioles on younger stems. Petioles 1–2 mm long, dark reddish to brown, smooth above, eglandular. Leaf blades (0.4–)

2.5–6.2 cm long, (1.0–) 1.8–3.7 cm wide, elliptic to ovate, base cuneate to rounded, apex obtuse (sometimes slightly acuminate), the tip rounded and rarely mucronate, margins flat; adaxial surface glabrous, oil glands common to abundant, visible with magnification; midvein flush; abaxial surface glabrous, oil glands few to common, visible with magnification; secondary veins indistinct to prominent; intramarginal veins indistinct to prominent, the inner intramarginal veins 1–3.5 mm from edge. Inflorescence terminal or axillary, solitary or paired; flowers solitary; anthopodia and metaxyphylls absent. Peduncles 12–28 mm long, rigid, glabrous to sparsely sericeous. Bracteoles 2, 1.3–2 mm long, 0.4–0.8 mm wide, linear to narrowly triangular, rigid, scalelike or occasionally foliaceous, midrib absent or occasionally present, with tips shorter than base of calyx lobes, glabrous to sparsely sericeous. Hypanthium 1.5–2 mm long, obconic, glabrous to sparsely sericeous, oil glands common. Calyx lobes 4, 1.7–3 mm long, broadly ovate, apex obtuse, greenish; adaxial surface sericeous to villous; abaxial surface glabrous to sparsely sericeous. Petals 4, 4.5–7 mm long, 4.2–5.5 mm wide, widely obovate; adaxial surface sparsely sericeous, white to yellowish white; abaxial surface sericeous, oil glands common. Stamens 65–120, included, filaments 4–5 mm long; staminal disk short-hairy; anthers 0.5–0.7 mm long, subglobose to globose, the tip of connective mostly eglandular or with a few glands. Style 4–5.5 mm long, glabrous to sparsely sericeous; ovary apex sparsely short-hairy; ovules 8–10 per placenta, placed irregularly. Berry reportedly black (Hyland & Whiffin 1993; Hyland et al. 1999). Seeds unknown. Fig. 20.

Phenology. Flowering from October through February; fruiting confirmed only for October.

Distribution (Fig. 16). Northeast Queensland from the McIlwraith Range south to near Cape Tribulation; locally common as an understory tree in rainforests and mesophyll vine forests on rocky slopes or ridges (Hyland & Whiffin 1993; Hyland et al. 1999); 200–600 m.

Conservation status. The species is considered Vulnerable (criterion C2a, Species Survival Commission 1994).

Additional Specimens Examined. **Australia.** Queensland. Cook District: Leo Ck, McIlwraith Ra., 13°40′S, 143°24′E, *Fell DF1415* (BRI); summit of Mt Cook, NP142, Cooktown, 15°29′S, 145°16′E, *Fell DF2427* (BRI, MEL, QRS); Mt Cook NP 142, 0.6 km NNW of summit, Cooktown, 15°29′S, 145°15′E, *Fell DGF2863* (BRI); Hope Vale Aboriginal Reserve, 5.7 km NE of Hope Vale Community, Cooktown, 15°16′S, 145°09′E, *Fell DGF3833* (BRI); Battle Camp rd, 1.4 km from McIvor–Isabella Rd, 15°16′S, 144°58′E, *Ford 2021* (BRI, QRS); Battle Camp rd, 1.4 km from Isabella–McIvor road, site 19, 15°16′S, 144°58′E, *Ford 2671 & Holmes, 2672* (BRI); 3 km NW of Isabella Falls on Battlecamp road to Laura, 15°17′E, 144°58′E, *Forster PIF14728* (BRI); South McIlwraith Ra., 13°58′S, 143°18′E, *Godwin C4108* (BRI); TR14, 13°45′S, 143°20′E, *Hyland 3322* (BRI); TR14, 13°45′S, 143°20′E, *Hyland RFK3259* (BRI); McIlwraith Ra., 13°50′S, 143°15′E, *Hyland 7630* (BRI); TR14, Leo Ck Rd, 13°43′S, 143°20′E, *Hyland 10126* (BRI, CANB, QRS); TR14, McIlwraith Ra., Leo Ck Rd, 13°45′S, 143°20′E, *Sanderson 779* (QRS); Upper Cameron Ck, NW of Cooktown, 15°22′S, 145°07′E, *Tracey 14167* (BRI); Emmagen Ck, N of Cape Tribulation, 16°03′S, 145°27′E, *Tracey 14757* (BRI, CANB, QRS); TR 14, 13°45′S, 143°20′E, Jul 1977, *Unwin 404* (QRS); headwaters of Massey Ck near old mining site, McIlwraith Ra., ca. 13°50′S, 143°20′E, *Webb & Tracey 9271* (BRI); headwaters of Lankelly Ck on western fall of McIlwraith Ra., ca. 13°52′S, 143°20′E, *Webb & Tracey 9574* (BRI).

The branchlets of *G. macilwraithensis* are often swollen prominently, and a pair of small bracts, or their remnant scars, is often located near the base of an internode on branchlets (e.g., *Tracey 14757*, BRI; *Unwin 404*, QRS). The 4-merous flowers sometimes have a few oil glands on the anther connective. The relative elongation of the leaf apex in *G. macilwraithensis* varies considerably. One specimen (*Unwin 404*, QRS) is aberrant in

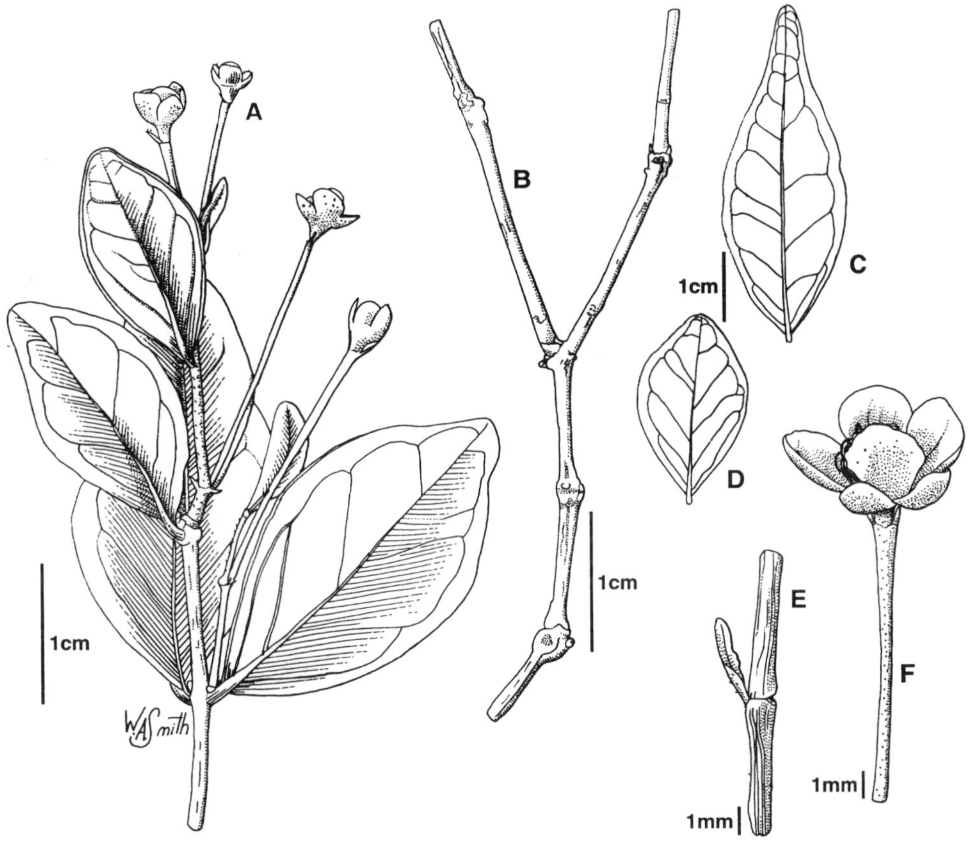

FIG. 20. *Gossia macilwraithensis*. A. Flowering branch. B. Branch with swollen nodes. C, D. Variation in leaf shape. E. Sterile bract near base of internode. F. Flower and peduncle. (Based on: A–C, F, *Hyland 11148*; D, E, *Goodwin 4108*.)

having slightly winged young internodes. The flowers are said to emit an unusual, unpleasant odor. The living bark is thin, whereas the dead bark is layered (Hyland & Whiffin 1993; Hyland 1999). The majority of herbarium material is sterile and fruits are unknown, which precludes a complete description at this time.

Gossia macilwraithensis resembles *G. lewisensis*, which has longer petioles and 5-merous flowers, and also *G. byrnesii*, which can have cordate leaf bases.

The epithet *macilwraithensis* refers to the primary geographical occurrence of the species in the McIlwraith Range.

17. Gossia bamagensis N. Snow & Guymer, sp. nov.—TYPE. AUSTRALIA. Queensland: Cook District, near Lockerbie, 10°47′S, 142°28′E, 1 Feb 1980, *B. Hyland 10235* (holotype: BRI!; isotypes: CANB! QRS!).

Ramuli rotundati; cortex laevis usque connspicue lamellatus; folia pagina supriore nitida glandibus oleosis vulgo visibilibus.

Trees, 5–30 m tall. Bark of trunk smooth, sometimes buttressed or fluted, sometimes conspicuously layered, reddish, organish, or somewhat greenish. Branchlets rounded to laterally compressed, not winged, light brown to brown, smooth or sometimes stringy, glabrous, oil glands common, indistinct to prominent. Leaves concolorous to slightly discolorous, somewhat glossy above, matte below. Stipules type A or type B (see Morphology). Petioles 2.5–7 mm long, yellowish to dark brown, smooth above to sometimes deeply channeled, eglandular. Leaf blades 3.0–12.0 cm long, 1.9–6.0 cm wide, elliptic to narrowly ovate, base cuneate, apex acute to acuminate, the tip rounded or pointed, margins flat; adaxial surface glabrous, oil glands common to abundant, visible with magnification; midvein flush; abaxial surface glabrous, oil glands common to abundant, prominent; secondary and intramarginal veins prominent; intramarginal veins 2, the innermost more prominent and ca. 1.5–2.5 mm from edge. Inflorescence terminal or axillary, solitary, paired or fascicled; flowers solitary or in racemes or panicles; anthopodia present or absent, metaxyphylls absent. Peduncles 1.5–16 mm long, rigid or drooping, glabrous to sparsely sericeous. Bracteoles mostly absent, sometimes 2, or a single enlarged bracteole, 1–1.8 mm long, 0.4–0.5 mm wide, linear to narrowly triangular, rigid, scalelike, midrib absent or faint, with tips shorter than base of calyx lobes, sparsely sericeous, persistent in fruit. Hypanthium 3–4 mm long, cupulate to obconic, sparsely sericeous near base, oil glands common. Calyx lobes 4, 2.5–3.5 mm long, broadly ovate, apex obtuse, greenish; adaxial surface sericeous to villous; abaxial surface glabrous but ciliate on margins. Petals 4, 4.5–6 mm long, 4.7–5.4 mm wide, oblate to circular, white to yellowish white; adaxial surface sparsely sericeous; abaxial surface somewhat sericeous, oil glands absent or sparse (or abundant, fide Hyland & Whiffin, 1993, and Hyland et al., 1999). Stamens up to 170, included, filaments 3.5–5 mm long; staminal disk glabrescent; anthers 0.5–0.7 mm long, subglobose to globose, the tip of anther connective eglandular. Style 4–4.5 mm long, sparsely sericeous below; ovary apex slightly hairy; ovules 10–12 per placenta, placed irregularly. Berry 9–11 mm long,10–13 mm in diameter, globose, base rounded, glabrous, dark purple to black. Seeds in mature fruit 1–3; more or less circular, flattened, smooth, dark brown, micropylar operculum absent, seed coat leathery and very thin. Embryo circinate, starchy, oil glands present, endosperm present; hypocotyl relatively thick, tip near radicle swollen, tip at same plane as embryo; cotyledons mostly straight; about equaling hypocotyl, pointing forward. Fig. 21.

Phenology. Flowering in February; fruiting in March.

Distribution (Fig. 12). Central and northeastern Queensland from near the northern tip of Cape York Peninsula near Bamaga to somewhat south of Prosperine; occasional to locally abundant in subcanopy or as emergents in rainforests, evergreen to deciduous notophyll vineforests, on flats, riverbanks, slopes, and upland areas on red, sandy or granitic soils; 10–800 m.

Conservation status. Although not well known, the species appears not to be endangered.

ADDITIONAL SPECIMENS EXAMINED. **Australia.** QUEENSLAND. COOK DISTRICT: King Park, Claudie R., between Ranger Base and river, 6.6 km WNW of Lockhart R. community, NE Cape York Peninsula, 12°46′S, 143°17′E, *Fell DGF3044* (BRI); upper catchment of Wilson Ck, Iron Ra. NP, 11.5 km N of Lockhart R. community, 12°41′01″S, 143°20′26″E, *Fell DGF 3112 & Butcher* (BRI, CANB); Round Mt, Embley Ra., 8.5 km WSW of the Nesbit R. mouth, 54.1 km NE of Coen, Silver Plains SLPF, Cape York Peninsula, 13°33′S, 143°30′E, *Fell DGF3433 et al.* (BRI); Hope Vale Aboriginal Reserve, 5.7 km NE of Hope Vale Community, 15°16′S, 145°09′E, *Fell DGF3832 & Stanton* (BRI); north footslopes of Ham Hill, 7.6 km NNW of Lockhart R., Departmental and Official Purposes Reserve D10, Cape Weymouth, 12°43′S, 143°19′E, *Fell DGF4124 &*

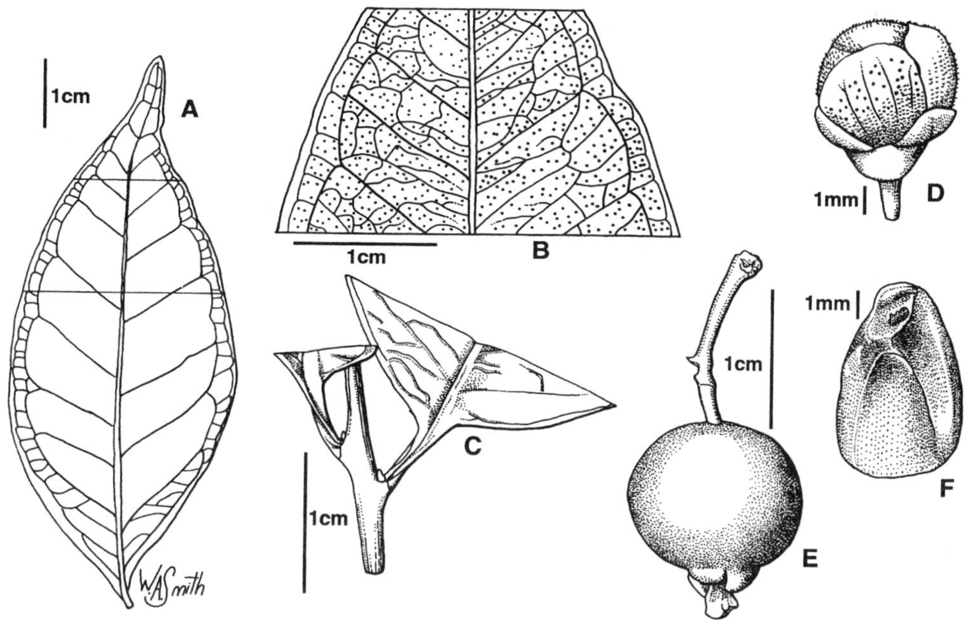

FIG. 21. *Gossia bamagensis.* A. Leaf. B. Detail of abaxial leaf surface. C. Node with a pair of leaves; note the channelled petioles. D. Flower bud. E. Fruit. F. Seed. (Based on: A, *Fell 3433*; B, *Volck 2630*; C, *Hyland 10235*; D, E, *Fell 4124*.)

Stanton (BRI); TR 14, Leo Ck mine area, McIlwraith Ra., 13°44′S, 143°22′E, *Forster PIF10209 et al.* (BRI); 10 km along Mt Lewis Rd, TR 66, Bushy LA, 16°35′S, 145°17′E, *Forster PIF15612 et al.* (BRI); Lockerbie Scrub, 10D46′S, 142°30′E, *Sankowsky 1208* (QRS); McIlwraith Ra., ca. 12 NE by E of Coeh, *Smith 11811* (BRI); north bank of Pascoe R. ca. 1 km from river mouth, 12°30′S, 143°15′E, *Tracey 14394* (QRS); Mt Webb, Starke Stn, 15°03′S, 145°05′E, *Tracey 14421* (QRS); Iron Ra., *Volck 2630* (BRI); Lockerbie Scrub, Bamaga, 10°48′S, 142°28′E, *Webb & Tracey Pen 18A* (QRS); Smithfield–Black Mt Rd, via Kuranda, *Webb & Tracey 7265* (BRI); McIlwraith Ra., NE of Coen, *Webb & Tracey 7338* (BRI); headwaters of Massey Ck near old mining site, ca. 13°50′S, 143°20′E, *Webb & Tracey 9143* (BRI); Rocky R. on eastern foothills of McIlwraith Ra., 13°47′S, 145°25′E, *Webb & Tracey 9444* (BRI, QRS).—North Kennedy District: Koolmoon Ck, 17°40′S, 145°30′E, *Smith 4730* (BRI).—South Kennedy District: Cawley SF, W of Cathu between Mackay and Proserpine, *Webb & Tracey 7672* (BRI).

Gossia bamagensis has the adaxial leaf surface generally a dull pale green, and somewhat waxy and shiny. The midvein of the abaxial surface is raised prominently near the petiole and often has prominent oil glands. Bracteoles are mostly absent, or are sometimes reduced to one or two small protrusions. This species is poorly known, and additional documentation of variation in floral and fruiting characters is desirable given the paucity of fertile material. Hyland et al. (1999) describe the bark as often conspicuously layered, but it is also reported as smooth (*Forster PIF13586 et al.*). The branchlets of a collection from a relatively high elevation in the Emble Range have somewhat stringy bark (*Fell DGF3433 et al.*, BRI).

The specific epithet refers to the Aboriginal settlement of Bamaga in northern Cape York Peninsula, within the geographical range of the species.

18. Gossia shepherdii (F. Mueller) N. Snow & Guymer, comb. nov. *Myrtus shepherdii* F. Mueller, Fragm. 9: 148. 1875. *Austromyrtus shepherdii* (F. Mueller) L. S. Smith, Proc. Roy. Soc. Queensland 67: 35. 1956.—TYPE: AUSTRALIA. Queensland: North Kennedy District, Rockingham's Bay, *J. Dallachy s.n.* (lectotype, here designated: MEL-92379!; isolectotypes: BRI! MEL-92377! MEL-92378!).

Trees, 3–15 m tall. Bark of trunk thin, smooth, often "python"-colored with shaded patches of browns, grays, reds, or greens, sometimes buttressed or fluted. Branchlets rounded to laterally compressed, not winged, light brown to light gray, smooth, glabrous, oil glands few to common, indistinct. Leaves concolorous to slightly discolorous, slightly glossy or matte above, matte below. Stipules type C (see Morphology). Petioles 3–8 mm long, smooth above, eglandular. Leaf blades 5.5–10.0 cm long, 2.0–5.7 cm wide, elliptic to broadly elliptic or sometimes ovate, base mostly cuneate or sometimes rounded, apex acuminate and sometimes resembling a drip tip, the tip rounded to pointed, margins flat; adaxial surface glabrous; oil glands few, visible only with magnification (if at all); midvein flush; abaxial surface glabrous, oil glands few to common, visible only with magnification (if at all); secondary and intramarginal veins prominent (the innermost more prominent), inner intramarginal vein 1.5–3.5 mm from edge. Inflorescence terminal or axillary, solitary or paired; flowers in racemes or rarely solitary; anthopodia and metaxyphylls absent. Peduncles 5–18 mm long, rigid, glabrous. Bracteoles 2, 1–1.7 mm long, 0.4–0.5 mm wide, linear to narrowly triangular, rigid, scalelike, midrib absent or rarely present, with tips shorter than base of calyx lobes, glabrous, persistent or caducous in fruit. Hypanthium 1.7–4.5 mm long, obconic, glabrous, oil glands common. Calyx lobes 4, 1.8–3 mm long, rounded, apex obtuse, greenish; adaxial surface sericeous to densely sericeous; abaxial surface glabrous but densely ciliate on margins. Petals 4, 4–9.5 mm long, 3–7 mm wide, obovate to oblate, white; adaxial surface glabrous to sparsely sericeous; abaxial surface glabrous, oil glands common. Stamens 95–155, included, filaments 4–10 mm long; staminal disk glabrous; anthers 0.5–0.9 mm long, subglobose to globose, the tip of connective eglandular. Style 3.5–8 mm long, glabrous; ovary apex glabrous; ovules 7–9 per placenta, placed irregularly. Berry 5–8 mm long, 5–8 mm in diameter, globose, base tapered, glabrous, dark blue-black. Seeds in mature fruit 1–4; rounded, smooth, ivory-colored, micropylar operculum absent, seed coat hard, bony. Embryo not seen.

Phenology. Flowering from August through March; fruiting August through March, peaking in September.

Distribution (Fig. 22). Northeastern Queensland; occasional to common in rainforest and evergreen simple notophyll vineforests; 650–1100 m.

Vernacular names. Lignum, shepherd's ironwood.

ADDITIONAL SPECIMENS EXAMINED. **Australia**. QUEENSLAND. COOK DISTRICT: Mt Windsor Tableland, 16°16′S, 145°04′S, *Crome 914* (BRI, QRS); SFR144, 16°16′S, 145°05′E, *Gray 407* (BRI, CANB, QRS), *412* (BRI, CANB, MO, QRS), *632* (BRI, CANB, QRS); SFR 144 Windsor Tableland, 16°15′S, 145°03′E, *Gray 1205* (BISH, BRI, CANB, MEL, MO, QRS); SFR144, 4.5 km along Trontero Ck Rd, SW of Millaa Millaa, 17°33′S, 145°34′E, *Guymer 2047* (BRI); Palmerston Track Ridge, 17°24′S, 145°45′E, *Hunter JH3834* (BRI); Lake Eacham NP, 17°16′S, 145°37′E, *Hunter JH4108* (BRI); SFR 1073, ca. 15 mi NW of Cairns, *Hyland 3256* (BRI); SFR144, 16°17′S, 145°05′E, *Hyland 3406RFK* (BRI, QRS); SFR 144, Whypalla, Bower Bird LA, 16°15′S, 145°01′E, *Hyland 25392* (BRI); 0.5 km past W Spencer Ck Forestry Camp, SF 144, Cockatoo LA, 33 km NNW Mt Carbine, 42 km NW of Mossman, Mt Windsor Tableland, 16°15′S, 145°02′E, *Jessup et al. GJM1198* (BRI); SFR 144, 0.4 km before W Spencer Ck Forestry Camp, Cockatoo LA, 0.3 km E Mt Carbine, Mt Windsor

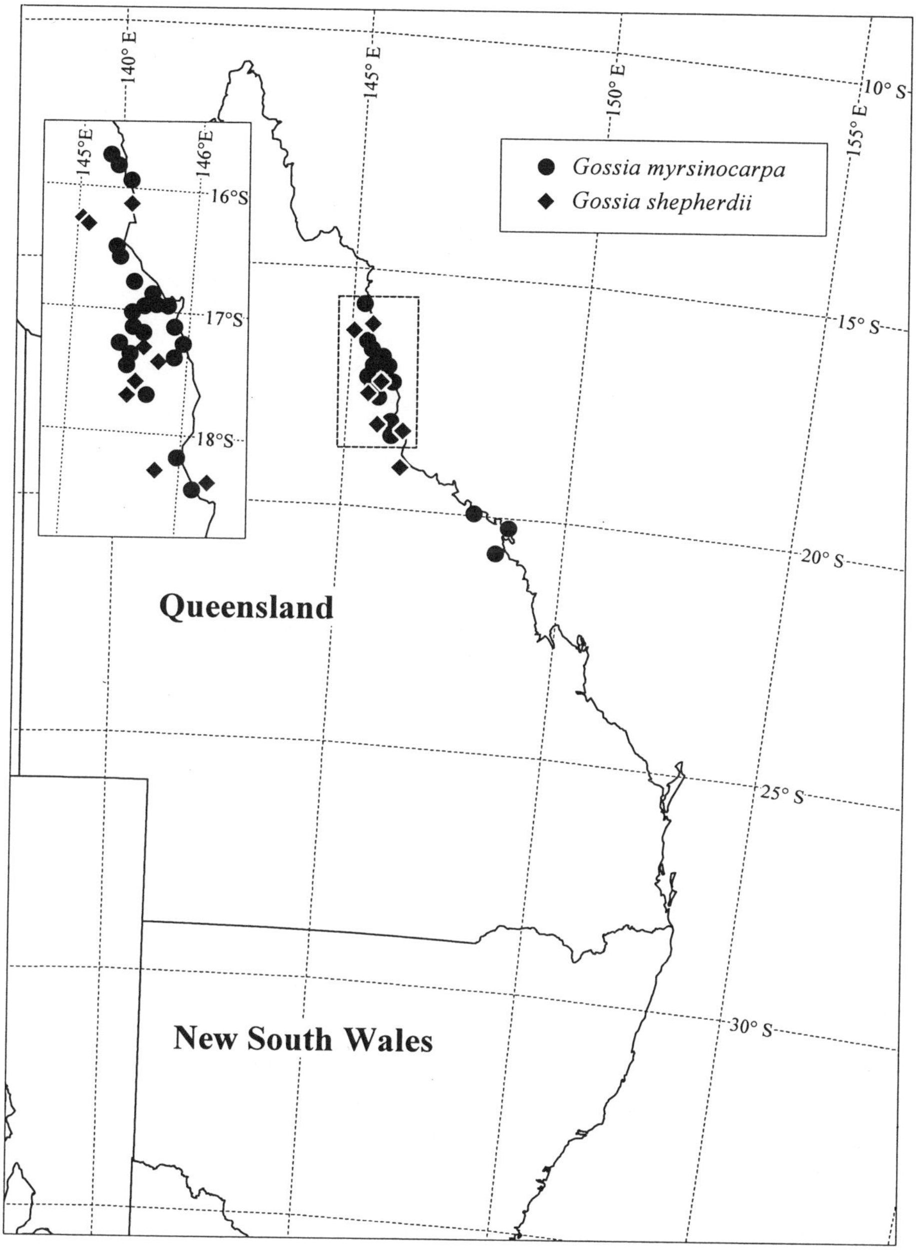

FIG. 22. Distribution of *Gossia myrsinocarpa* and *G. shepherdii*.

Tableland, 16°15′S, 145°02′E, *Jessup et al. GHM1241* (BRI); Mt Bartle Frere track, Boonjee LA, near Bobbin Bobbin Falls, 17°22′S, 145°46′E, *Jessup et al. GJM5214* (BRI); SFR 144, 16°15′S, 145°02′E, *Moriarty 2622* (QRS); Atherton Tableland, *Tardent* 231 (BRI); Mt Hemmant just N of Noah Ck in Cape Tribulation area, 16°06′S, 145°26′E, *Webb & Tracey 11709* (BRI, CANB, GREE, QRS); Malanda, Jan 1923, *White s.n.* (BRI 319176, 319177).—NORTH KENNEDY DISTRICT: Saddle below main ridge of Mt Bowen, Hinchinbrook Is., 18°21′S, 146°14′E, *Fell DF1212 & Swain* (BRI); SFR 605, 18.8 km along Culpa road from Koombooloomba rd, site 43, 17°58′S, 145°38′E, *Ford et al. 3176* (GREE); Hinchinbrook Is., main northern creek of Little Ramsay Bay, headwaters of creek, 18°S, 146°E, Aug 1975, *Hockings s.n.* (BRI 250142); Mt Spec, 19°00′S, 146°05′E, *Hyland 1562* (BRI); Dotswood Holding, 19°10′S, 146°22′E, *Hyland 10089* (BRI, CANB); N slope of Mt Thorn, SF 461 Meunga LA, 23.1 km W of Cardwell, site 5, perimeter, 18°16′S, 145°47′E, *Jessup et al. GJM2295* (BRI); Koolmoon Ck, 17°40′S, 145°30′E, *Smith 4741* (BRI); Mt Spec, ca. 6 km S of Ingham, 18°57′S, 146°12′E, *Smith 5352* (BRI, CANB, RSA); Mt Spec, *White 905* (BRI).

The bark of *G. shepherdii* is smooth and often has "python" coloration. The species most closely resembles *G. acmenoides* and *G. bidwillii*, but these differ in having 5-merous flowers. The abaxial leaf surface lacks oil glands or bears minute ones that are seen only with magnification. Seeds bearing mature embryos have not yet been found (this study; Hyland & Whiffin 1993; Hyland et al. 1999). The flowers are reported to be heavily perfumed (*Gray 1205*).

The reference specimen of *Austromyrtus* sp. "Windsor Tableland" (*Gray 412*, as reported in Henderson, 1997, and Hyland et al.,1999) is virtually identical to the lectotype of *G. shepherdii* with respect to characters of the leaves, flowers, and stems. The only feature showing variation is the degree of hairiness of the upper surface of the calyx lobes, which varies from sparsely to densely sericeous.

The essential oil profiles (Brophy et al. 1996a) of the voucher specimen of *G. shepherdii* (*Forster PIF13671*, BRI) are similar in some notable respects to those of a specimen (*Forster PIF15535 et al.*, BRI) that was originally assigned to the provisionally segregated taxon "Windsor Tableland" (as "Pinnacle Track" in Brophy et al. 1996a). Both have relatively high amounts of bicyclogermacrene (15.25% and 10.73%, respectively), which otherwise is only found in high concentrations in *G. macilwraithensis* (16.69%); globulol (9.02% and 7.68%); and viridiflorol (4.63% and 4.72%). The two collections, however, differed in other respects. *Forster PIF15535 et al.* had the following oils in excess of 3%, which were absent or in very low amounts (< 1%) in *Forster PIF13671*: α-pinene (3.62%), ß-pinene (13.95%); aromadendrene (3.60%), α-terpineol (3.04%). The voucher of *G. shepherdii* had 3% or more of the following oils, which were absent in *Forster PIF15535 et al.*: δ-elemene (3.80%); γ-elemene (3.35%); viridiflorene (5.63%); germacrene-B (19.62%). The similarity of oil profiles does not contradict assigning the specimen designated as "Windsor Tableland" to *G. shepherdii.*

Most likely this species was named for Thomas William Shepherd (1824–1884), son of Thomas Shepherd (1779–1835), a Scottish nurseryman who established the Darling Nursery in 1827.

19. Gossia lucida (Banks ex Gaertner) N. Snow & Guymer, comb. nov. *Syzygium lucidum* Banks ex Gaertner, Fruct. sem pl. 1: 167, t. 33. 1788. *Myrtus nitida* J. F. Gmelin, Syst. veg. 2: 792. 1791, non *Myrtus lucida* L., 1753. *Austromyrtus lucida* (Banks ex Gaertner) L. S. Smith, Proc. Roy. Soc. Queensland 67: 35. 1956.—TYPE: AUSTRALIA. "New Holland," 1770, *Banks & Solander s.n.* (holotype: K!, photo: BRI!).

Myrtus monosperma F. Mueller, Vict. Nat. 9: 9. 1892.—TYPE: AUSTRALIA. Queensland. Cook District, Endeavour R., *W. A. Persieh s.n.* (holotype: MEL!; isotypes: BM! BRI! MEL!).

Multi-stemmed shrubs to trees, 5–8 m tall. Bark of trunk smooth, occasionally stringy or fluted, variously brown or gray and sometimes blotched with patches of orange or green. Branchlets rounded, not winged, light brown to light gray, smooth, glabrous, oil glands absent to indistinct but common. Leaves slightly discolorous to discolorous, glossy or matte above, matte below. Stipules type B (see Morphology) but usually obscure. Petioles 2–3.5 mm long, dark brown, smooth above, eglandular. Leaf blades 5.0–10.0 cm long, 2.5–7.0 cm wide, elliptic to broadly elliptic or narrowly ovate, base cuneate to rounded, apex acute to acuminate, the tip rounded, margins flat; adaxial surface glabrous, oil glands abundant, prominent (but smaller than in other most other species); midvein flush; abaxial surface glabrous, oil glands abundant, prominent but small; secondary and intramarginal veins prominent, the inner intramarginal vein 1–4 mm from edge. Inflorescence terminal or axillary, solitary, paired, or fascicled; flowers solitary or in racemes or panicles; anthopodia present (up to 2 mm long) or absent, metaxyphylls present (sometimes disjunct opposite) or absent. Peduncles 10–22 mm long, rigid, glabrous. Bracteoles 2 (usually falling early), 1–1.5 mm long, 0.3–0.5 mm wide, linear, rigid, scalelike, midrib absent, the tips shorter than base of calyx lobes, glabrous, caducous in fruit. Hypanthium 1.3–2 mm long, obconic to cupulate, glabrous, oil glands few to common. Calyx lobes 4, 1.5–2 mm long, broadly triangular to rounded, apex obtuse, greenish or rarely petaloid; adaxial surface sparsely sericeous to sparsely villous; abaxial surface glabrous. Petals 4, 6.0–6.5 mm long, 4.5–5 mm wide, widely ovate to obovate or oblate, yellowish white; adaxial surface glabrous but slightly ciliate on margins; abaxial surface glabrous, oil glands few to common. Stamens 75–100, included, filaments 3–5 mm long; staminal disk short-villous; anthers 0.5–0.6 mm long, subglobose, the tip of connective with a single orange gland. Style 4–5 mm long (sometimes geniculate near apex when dry), glabrous; ovary apex glabrous; ovules up to 11 per placenta, placed irregularly. Berry 6.5–10 mm long, 6–10 mm in diameter, subglobose, base rounded, glabrous, red to dark blue or black (at maturity). Seeds in mature fruit 1–2; rounded to subglobose, compressed laterally, smooth to rough, light brown, micropylar operculum absent, seed coat hard, bony. Embryo circinate, starchy, oil glands present or absent, endosperm scant but apparently present (needs confirmation); hypocotyl relatively thick, tip near radicle swollen, tip at same plane as embryo; cotyledons mostly straight, shorter than hypocotyl, pointing forward.

Phenology. Flowering March through May; fruiting July through March.

Distribution (Fig. 9). Northeastern Queensland; along river banks to ridges in mesophyll or notophyll vineforests on granites or metamorphosed substrates; 10–1000 m.

Conservation status. The species should be considered Vulnerable (criterion D1, Species Survival Commission 1994); future reports may suggest that subpopulations are more extensive than currently known.

ADDITIONAL SPECIMENS EXAMINED. **Australia**. QUEENSLAND. COOK DISTRICT: Mt Cook NP NP#142, 0.6 km NNW of summit, Cooktown, 15°29′S, 145°15′E, *Fell DGF2848A* (BRI); Cape Melville NP NP4, Eumangin/Temple Ck catchment, 11.5 km NW of Barrow Point, 79.2 km NNE of Lakefield Ranger Base, 14°17′S, 144°32′E, *Fell DGF4317* (BRI); Bathrust Range, 19 km SSE of Bathurst Head, 55.8 km N of Lakefield Ranger Base, 14°25′S, 144°15′E, *Fell DGF4498* (BRI); NPR 166, Black Mountain, Helenvale road, site 17, 15°40′S, 145°13′E, *Ford AF2645 & Holmes* (BRI, CANB, GREE, NSW); Kennedy Hill Gorge, 12°28′S, 143°16′E, *Forster et al. PIF5413* (BRI); 1 km E of Kennedy Hill, facing sea, 12°28′S, 143°16′E, *Forster PIF5420 & Tucker* (BRI, MEL, MO); SF 185 Danbulla, 0.5 km down Kauri Cr (B) Rd, 17°05′S, 143°35′E, *Forster et al. PIF15604* (BRI); Portion 17V, Parish of Monkhouse, 15°42′S, 145°13′E, *Hyland 14035* (QRS); Annan R.–Cooktown Developmental Rd crossing, north bank, crossing closest to Cooktown, 15°30′S, 145°10′E, *Irvine*

596 (BRI, CANB); 6.5 km S of McIvor R. crossing on McIvor R. Rd, 800 m W of road in sheltered area along creek, 15°11′S, 145°08′E, *Jessup et al. GJD3055* (BRI); Mt Lewis, North Mary LA R 143, E/P 18, 16°30′S, 145°16′E, *Sanderson 481* (QRS); Main Coast Ra., 16°22′S, 145°18′E, *Sankowsky 1273 & Sankowsky* (BRI); Annan R. crossing, ca. 8 km SSW of Cooktown, 15°41′S, 145°13′E, *Smith 10651* (BRI, CANB); Gap Ck, ca. 22 mi SE of Cooktown, 15°S, 145°E, *Smith 14414* (BRI); 2 km from Cooktown along McIvor R. road, 15°29′S, 145°14′E, *Telford 9395 & Butler* (BRI, CANB); McIvor R. crossing, Cooktown–Starke Rd, 15°07′S, 145°08′E, *Tracey 14434* (BRI, CANB, MO, QRS); Boonjie LA on the Atherton Tableland, 17°S, 145°E, *Tucker MCT85* (BRI); Dowlings Hill on Mt Amos Rd, S of Cooktown, 15°38′S, 145°18′E, *Webb & Tracey 11881* (BRI); Mt Bourgamba between Gold Hill near China Camp and Alexandra Ck, 16°07′S, 145°19′E, *Webb & Tracey 13733* (BRI, QRS); Endeavour R., unknown collector (MEL 23859, 238760, 238761).—NORTH KENNEDY DISTRICT: Palm Islands, *Bancroft 99* (BRI).—SOUTH KENNEDY DISTRICT: Mt Beatrice NP, headwaters of Catherine Ck, 20°52′S, 148°40′E, *Forster PIF14662* (BRI).

CULTIVATED: Tolga, collected near Hopevale, N of Cooktown, 17°13′S, 145°29′E, *Sankowsky 1579* (BRI).

The smooth, often multicolored bark is partially diagnostic of *G. lucida*. The berries at maturity are red or dark purplish. In contrast to *G. shepherdii*, which also has smooth bark and overlaps geographically with *G. lucida*, the oil glands of the leaf are abundant and usually visible without magnification. The apex of the hypanthium is often reddish (A. Ford, pers. comm., 2001).

Banks's choice of the epithet *lucida* ("bright, shining") probably refers to the glossy foliage.

20. Gossia floribunda (A. J. Scott) N. Snow & Guymer, comb. nov. *Backhousia floribunda* A. J. Scott, Kew Bull. 39: 659. 1984. *Austromyrtus floribunda* (A. J. Scott) Guymer, Austrobaileya 2: 569. 1988.—TYPE: PAPUA NEW GUINEA. Kwinia Ck, near Ellangowan Is., middle Fly River, 21 Oct 1967, *R. Pullen 7510* (holotype: K!; isotypes: A! BO, BRI! CANB! G, L! LAE, PNH, US).

Shrubs to trees, 2–15 m tall. Bark of trunk smooth, mostly brown or whitish but often with patches of various colors. Branchlets distinctly 4-angled but sometimes laterally compressed, edges distinctly winged (wings up to 1 mm wide), whitish to reddish brown, smooth to slightly flaky, sericeous but soon becoming glabrous, oil glands absent to common but indistinct. Leaves slightly discolorous to discolorous, glossy or matte above, matte below. Stipules type B (see Morphology). Petioles 1.5–4.5 mm long, dark brown to reddish, channeled, eglandular. Leaf blades 1.5–7.5 cm long, 1.0–4.0 cm wide, elliptic to narrowly ovate, base cuneate, apex acute to acuminate, the tip rounded to pointed or mucronate, margins flat; adaxial surface glabrous, oil glands few to common, barely visible to prominent; midvein raised; abaxial surface glabrous, oil glands few to abundant, barely visible to prominent; secondary and intramarginal veins prominent or indistinct, the inner intramarginal vein 0.5–1.5 mm from edge. Inflorescence terminal or axillary, a monad on hairy short shoots or in racemes; anthopodia mostly absent, metaxyphylls absent. Peduncles 4–15 mm long, rigid to flexuous, glabrous. Bracteoles 2 and flanked by large hairlike "stipules," 0.5–1.7 mm long, ca. 0.3 mm wide, linear, rigid, scalelike, midrib absent, with tips shorter than base of calyx lobes, sparsely sericeous, persistent or caducous in fruit. Hypanthium 1.5–2.5 mm long, cupulate, glabrous to sparsely sericeous, oil glands few to abundant. Calyx lobes 4 (rarely 5), 1–2 mm long, broadly triangular to rounded, apex obtuse, greenish or somewhat petaloid; adaxial surface glabrous but sparsely ciliate on margins; abaxial surface glabrous. Petals 4 (rarely 5), 2–6 mm long, 2–4 mm wide, ovate to obovate, white; adaxial surface glabrous but ciliate on margins; abaxial surface glabrous, oil glands common but faint. Stamens 55–85, included, filaments 1.5–6 mm long; staminal disk glabrous to glabrescent; anthers 0.4–0.6 mm long, subglobose to globose, tip of connectives eglandular. Style 2.5–5.5 mm long, glabrous; ovary apex glabrous to sparsely hairy; ovules 5–16 per placenta, placed irregularly. Berry 6–11.5 mm long, 7–11 mm in diameter, subglobose to globose, base rounded, glabrous, grayish to purple or black. Seeds in mature fruit 1–2, rounded to suborbicular, smooth, greenish brown, micropylar operculum absent, seed coat hard and bony but thin. Embryo circinate, starchy, oil glands present or absent, endosperm present but scant; hypocotyl relatively thick, tip near radicle swollen, at same plane as embryo; cotyledons circinate or straight, shorter than hypocotyl, folded back towards hypocotyl or pointing forward.

Phenology. Flowering November through January; fruiting mostly February through June.

Distribution (Fig. 13). Southern Papua New Guinea to northeastern Queensland; uncommon to very common in rainforest, evergreen or semideciduous notophyll vineforest, or vine thickets; sea level to 800 m.

Vernacular names. Cape ironwood, lignum.

ADDITIONAL SPECIMENS EXAMINED. **Australia.** QUEENSLAND. COOK DISTRICT: 9.0 km N of Batavia Downs on the Peninsula Development Rd, 12°35′S, 142°40′E, *Clarkson 8510 & Nelder* (BRI); just W of Lake Wicheura 4.5 km W of Vallack Point, 22.6 km NE of Bamaga, 10°46′S, 142°33′E, *Fell et al. DGF3891* (BRI); Frangipanni Beach Scrub 0.7 km S of Cape York 26.4 km NE of Bamaga, 10°41′S, 142°31′E, *Fell et al. DGF3999* (BRI); 12.6 km WNW of Lockhart R. community, 12°44′S, 143°14′E, *Fell DGF4201 & Stanton* (BRI); McIvor–Isabella road, 3.1 km from Battle Camp road, site 20, 15°15′S, 144°59′E, *Ford AF2674 & Holmes* (BRI, GREE); Lockerbie Scrub, 3 km past Lockerbie HS, 10°47′S, 142°29′E, *Forster PIF6317* (BRI); Wenlock R., S of Stone Crossing, 12°23′42″E, 142°10′59″S, *Forster PIF14392* (BRI); Newcastle Bay, stabilised dune behind headland between Nanthau and Narau beaches, 10°46′S, 142°35′E, *Forster PIF15299 & Tucker* (BRI); CSIRO-QRS Arboretum, Atherton, *Gray 2237* (BRI, QRS); Bamaga Rd 8 km S of Cape York, 10°45′S, 142°31′E, *Gray 4275* (BRI); 4 km N of Lockerbie, 10°47′S, 142°29′E, *Gray 4281* (BRI); CSIRO FRS, Atherton 17°15′S, 145°30′E, *Hyland 9629* (BRI, QRS); Massy Ck, 13°55′S, 143°30′E, *Hyland 10289* (BRI, CANB); TR 14 Massy, 13°52′S, 143°25′S, *Hyland 10869* (BRI); Claudie R., 12°44′S, 143°15′E, *Hyland 13102* (BRI); 4 km W of Isabella Falls on Battle Camp Rd, 15°16′S, 144°59′E, *Jessup et al. GJD3024* (BRI); Lockerbie, ca. 20 km N of Bamaga on Cape York Rd, 10°48′S, 142°28′E, *Jones 2265* (BRI); CSIRO-QRS Arboretum, Atherton, *Nella 2* ["peg number" 628] (BRI, CANB, QRS); Tolga, 17°13′S, 145°29′E, *Sankowsky 1102 & Sankowsky, 1301, 1309, 1364, 1408* (BRI); CSIRO-QRS Arboretum, Atherton, *Starkey 50* ["peg number" 628] (BRI).

CULTIVATED: Cultivated at CSIRO Atherton, seed originally from Claudie R., 17°17′, 145°29′E, *Guymer 2052* (BRI, GREE, MEL).

The quadrangular branchlets, the prominent, basally swollen stipular hairs, and the raised midvein on the adaxial leaf surface are diagnostic. Flowers are either 4- or 5-merous and can be strongly scented.

The specific epithet alludes to the relative abundance of flowers in the inflorescence.

21. Gossia myrsinocarpa (F. Mueller) N. Snow & Guymer, comb. nov. *Eugenia myrsinocarpa* F. Mueller, Victorian Nat. 8: 200. 1892.—TYPE: AUSTRALIA. Queensland: Cook District, Trinity Bay, *Fitzalan s.n.* (holotype: MEL; isotype: MEL).

Austromyrtus minutiflora Burret, Notizbl. Bot. Gart. Berlin-Dahlem 15: 502. 1941.—TYPE: AUSTRALIA. Queensland: North Kennedy District, Rockingham Bay, collector unknown (holotype: B, destroyed; isotype: MEL!).

Shrubs or trees, 2.5–15 m tall. Bark of trunk rough, often scaly, fissured, or tessellated, brown to gray. Branchlets rounded, not winged, light gray, smooth, glabrous, oil glands indistinct to prominent, common. Leaves concolorous to discolorous, matte above and below. Stipules type C (see Morphology) but obscure. Petioles 3–7 mm long, green to reddish or dark brown, usually slightly channeled, eglandular. Leaf blades 4.5–8.2 cm long, 1.7–3.6 cm wide, elliptic to narrowly ovate, base cuneate, apex acute to acuminate, the tip rounded to pointed, margins flat; adaxial surface glabrous, oil glands abundant, visible with magnification; midvein flush; abaxial surface glabrous, oil glands abundant, visible with magnification; secondary veins prominent; intramarginal veins indistinct to prominent, the inner intramarginal vein 1.5–3 mm from edge. Inflorescence terminal or axillary, solitary, paired, or fascicled; flowers solitary, or in triads or racemes; anthopodia absent or occasionally present, metaxyphylls present or absent. Peduncles 3–11 mm long, rigid, glabrous. Bracteoles absent or sometimes 2, 0.8–1.5 mm long, 0.3–0.4 mm wide, linear to narrowly triangular, rigid, scalelike without midrib, the tips shorter than base of calyx lobes, glabrous, persistent or caducous in fruit. Hypanthium 1.0–2.5 mm long, cupulate, glabrous, oil glands abundant. Calyx lobes 4, 0.7–2.5 mm long, broadly ovate, apex obtuse, greenish or somewhat petaloid; adaxial surface sparsely sericeous near margins; abaxial surface glabrous. Petals 4, 2.5–5.5 mm long, 2–2.8 mm wide, ovate to widely obovate or circular, white to yellowish white; adaxial surface glabrous but ciliate on margins; abaxial surface glabrous, oil glands common. Stamens 55–75, exserted (fide Hyland et al., 1999, see photo), filaments 2–4.5 mm long; staminal disk glabrous to sparsely short-hairy; anthers 0.4–0.6 mm long, subglobose to globose, the tip of connective with a single gland. Style 2–4.5 mm long, glabrous to sparsely sericeous; ovary apex glabrous to hairy; ovules 10–12 per placenta, placed irregularly. Berry 5–11 mm long, 5–12 mm in diameter, subglobose to globose, base rounded or rarely tapered, glabrous, dark blue-black. Seeds in mature fruit 1–2; semicircular, somewhat flattened, smooth, ivory-colored, micropylar operculum absent or occasionally present, seed coat hard, bony. Embryo circinate, starchy, oil glands present but faint, endosperm present but scant; hypocotyl relatively thick, tip near radicle swollen but thickest subapically, tip at same plane as embryo; cotyledons circinate, shorter than hypocotyl, pointing forward.

Phenology. Flowering occurs from December through February; fruiting December through September.

Distribution (Fig. 22). Central and northeastern Queensland, from somewhat south of Cooktown to the Proserpine area; locally common on hillsides and streamsides in mesophyll or notophyll rainforest, vineforests, and palm forests; 5–1000 m.

Vernacular name. Malanda ironwood.

ADDITIONAL SPECIMENS EXAMINED. **Australia.** QUEENSLAND. COOK DISTRICT: Rex Ra. NE of Julatten, 16°34′S, 145°23′E, *Bean 5680* (BRI); Rockingham Bay, May 1865, *Dallachy s.n.* (MEL 92359); bank of Campbells' Ck, 17°06′S, 145°52′E, *Flecker 3313* (QRS), *6954* (QRS); Clayton's Ck, 16°56′S, 145°48′E, *Flecker 8851* (QRS); Cairns Intake, Freshwater Ck, 16°55′S, 145°42′E, *Flecker 10795* (QRS); SF 191, Wongabel, 17°19′S, 145°30′E, *Forster PIF14443* (BRI); 17°19′30″S, 145°31′02″E, *Forster PIF17161 & Figg* (BRI); SF 191, Wongabel, 17°20′S, 145°30′E, *Gray 233* (BRI, CANB, QRS); SFR194, Crater, 17°26′S, 145°29′E, *Gray 895* (BRI, CANB, MEL, MO, QRS); SFR 191, 17°20′S, 145°30′E, *Gray 1674* (BRI, CANB, MO, QRS); SFR 675, Parish of Cairns, Shoteel LA, 16°55′S, 145°36′S, *Gray 3100* (BRI), *3272* (BRI, QRS); NP Reserve, Parish of Alexandra, Heights of Alexandra, 16°15′S, 145°26′E, *Gray 3809* (BRI); SFR 607, Bridle LA, 16°58′S, 145°35E, *Gray 5999* (BRI, QRS); Mossman Gorge NP, 16°28′S, 145°20′E, *Guymer 2040* (BRI, CANB); Barron Falls Track, 16°51′S, 145°39′E, *Guymer 2054* (BRI); Tinaroo Ra., ca. 9 km NE of Tinaroo Dam, 17°09′S, 145°35′E, *Guymer 2072* (BRI); Ridings LA, SFR185, 17°08′S, 145°36′E, *Irvine 1809* (BRI, CANB, QRS); SFR 1073, 16°45′S, 145°30′E, *Hyland 2303* (BRI); NPR 904, 17°40′S, 145°40′E, *Hyland 3970* (BRI); Public Purposes Reserve 277, Sophia, 17°14′S, 145°57′E, *Hyland 9345* (BRI, CANB, QRS); SF 194, Parish of Western, 17°15′S, 145°25′E, *Hyland 15048* (QRS); The Boulders, Babinda, 17°21′S, 145°53′E, *Jago 389* (QRS); Rex Ra. ca. 2.8

km from Mossman–Julatten road on forest track, ca. 9 km NE of Julatten, 16°32′S, 145°55′E, *Jessup et al. 875* (BRI); top of the Gap, N of Mt Finnigan, just S of Rossville along roadside, 15°48′S, 145°18′E, *Jessup et al. GJD3115* (BRI); Gap Ck, Bloomfield, 15°S, 145°E, *Johnson 1401A* (BRI); Gadgarrah, Atherton, *Kajewski 1053* (BRI); Lake Barrine, Atherton Tableland, *Kajewski 1343* (BRI, NY, SING); Mourilyan Harbour, Feb 1912, *Kenny s.n.* (BRI AQ093498); foot of MacAlister Ra., 2.5 km ENE of Saddle Mt, 16°49′S, 145°41′E, *Lyons 55* (BRI); SFR 185, Danbulla, Downfall LA, 17°10′S, 145°37′E, *O'Farrell 110* (BRI, CANB, QRS); Wongabel SF 191, 17°20′S, 145°35′E, *Risley 32* (BISH, BRI, CANB, NSW, QRS, SING); northeast margin of Lake Tinaroo, ca. 6 mi NE of Tinaroo, ca. 17°07′S, 145°31′E, *Schodde 3300* (BRI, CANB); Gap Ck, ca. 38 km S by E of Cooktown (9 km by road from Rossville),15°43′S, 145°14′E, *Smith 10729* (BRI); Gap Ck, ca. 22 mi SE of Cooktown, 15°S, 145°E, *Smith 14428* (BRI, CANB); Freshwater Ck, adjacent to Por. 38V, Cairns, 7 km S of Redlynch, 16°S, 145°E, *Stanton 1022* (BRI); Kuranda, *Webb & Tracey 5707* (BRI, CANB, GREE, K, MEL, MO); Danbulla, Mobo Ck Rd, 17°0′S, 145°30′E, *Webb & Tracey 5730* (BRI, GREE, K, MEL, MO); Mt Spurgeon, *White 10613* (BRI); Mossman Gorge, W of Mossman, *Wrigley & Telford 1159* (BRI, CANB).—NORTH KENNEDY DISTRICT: Edmund Kennedy NP, near Cardwell, 18°10′S, 145°58′E, *Bean 3899* (BRI, MEL); Impulse Ck, SF 299, Conway, 20°12′S, 148°44′E, *Forster PIF8292 & McDonald* (BRI, MEL); Muller Ck Conn LA, 21 km S of Cardwell, 18°25′S, 146°07′E, *Jessup et al. GJM2483* (BRI); Brandy Ck Rd, Conway SF, 20°20′S, 148°41′E, *Perry 2* (BRI), Aug 1990, *Perry s.n.* (BRI AQ473230); Townsville Rd, 14 km N of Bowen, on W side of road 100 m N of Euri Ck, 19°27′S, 147°57′E, *Sharpe 84* (BRI); Kirrama Ra., W of Kennedy, *Smith 3211A* (BRI); Kirrama Ra., W of Kennedy, *Smith 3215* (BRI).—SOUTH KENNEDY DISTRICT: crest of Clark Ra., SF 652 Cauley, ca. 15 km SW of Bloomsbury, 20°43′S, 148°26′E, *McDonald 5997 & Champion* (BRI); Cathu SFR, W of Sandy Ck area, 20°49′S, 148°38′E, *Pearson SP561* (BRI).

CULTIVATED: Tolga (cult.), *Sankowsky 1554* (QRS); in Tolga ex Mt Lewis (near hut), *Sankowsky 1537* (QRS).

The small, 4-merous flowers of *G. myrsinocarpa* are the most reliable diagnostic character, but the relatively distinct intramarginal veins are helpful for determining sterile specimens. More so than in other species in the genus, young leaves of *G. myrsinocarpa* are often disjunct opposite on developing branchlets (e.g., *Risley 508*, QRS, *Sanderson 677*, QRS), which is also helpful for identifying sterile material. Many specimens (e.g., *Gray 233, 3809*, BRI) have an even number of flowers at the apex of botryoid-like inflorescences rather than the typical triad (central flower with two lateral flowers). *Gossia myrsinocarpa* resembles *G. bidwillii*, but that species has 5-merous flowers, larger oil glands on the leaves, and more deeply channeled petioles.

The epithet *myrsinocarpa* is derived from the Greek *myrsine* (myrtle) and *karpos* (fruit).

22. Gossia fragrantissima (F. Mueller ex Bentham) N. Snow & Guymer, comb. nov. *Myrtus fragrantissima* F. Mueller ex Bentham, Fl. austral. 3: 277. 1867. *Austromyrtus fragrantissima* (F. Mueller ex Bentham) Burret, Notizbl. Bot. Gart. Berlin-Dahlem 15: 501. 1941.—TYPE: AUSTRALIA. Queensland/New South Wales: Moreton Bay and Richmond R., *C. Moore s.n.* (holotype: K! photo: BRI!; isotype: MEL!).

Large shrubs to trees, 2–10 m tall. Bark of trunk rough, flaky, or fissured, brown. Branchlets rounded to laterally compressed, not winged, brown to reddish brown, smooth, puberulous, the hairs less than 0.5 mm long, oil glands sparse, indistinct. Leaves slightly discolorous, glossy above, matte below. Stipules type A (see Morphology), hairs whitish or ferrugineous. Petioles 1.5–4 mm long, green to brown, channeled, eglandular. Leaf blades 14–5.6 cm long, 0.8–3.1 cm wide, narrowly elliptic to ovate or obovate, base cuneate, apex mostly acute, the tip mucronate or rarely retuse, margins flat to undulate; adaxial surface glabrous, oil glands common, visible without magnification; midvein flush; abaxial surface glabrous but sparsely sericeous on midvein, oil glands common, usually visible only with magnification; secondary veins prominent; intramarginal veins indistinct to prominent, the inner intramarginal vein 0.5–1.5 mm from edge. Inflorescence axillary, solitary; flower solitary or in a raceme; anthopodia and metaxyphylls absent. Peduncles rigid, 4–6 mm long, short-sericeous. Bracteoles 2, 1–2 mm long, 0.2–0.5 mm wide, linear to narrowly triangular, rigid, scalelike without midrib, with tips shorter than base of calyx lobes, shortly sericeous, persistent or caducous in fruit. Hypanthium 1.5–1.8 mm long, obconic to cupulate, short-sericeous throughout, oil glands common. Calyx lobes 4, 1–2 mm long, broadly triangular, apex acute, generally petaloid; adaxial surface sparsely sericeous (especially near apex); abaxial surface sparsely sericeous. Petals 4, 2.5–4 mm long, 3–3.5 mm wide, ovate to widely ovate, white, often strongly reflexed at anthesis; adaxial surface glabrous; abaxial surface sericeous throughout, oil glands absent. Stamens ca. 80, excluded, filaments 2–3 mm long; staminal disk short-hairy; anthers 0.5–0.7 mm long, subglobose to globose, the tip of connective eglandular or with a small faint gland. Style ca. 4 mm long, glabrous or sparsely sericeous below; ovary apex glabrous; ovules 8–10 per placenta, placed irregularly. Berry 6–7.5 mm long, ca. 7.5 mm in diameter, globose, base rounded, glabrous, orangish yellow to red at maturity. Seeds in mature fruit 2, more or less circular but flattened, smooth, light brown or ivory-colored, micropylar operculum present, seed coat hard, bony. Embryo circinate, oily on surface, oil glands on embryo present, endosperm absent; hypocotyl relatively thick, tip near radicle not or barely swollen, tip at same plane as embryo; cotyledons circinate, shorter than hypocotyl, pointing forward.

Phenology. Flowering from September through November; fruiting November through March.

Distribution (Fig. 18). Northern New South Wales and southeastern Queensland; on ridges, slopes, and riverine sites in rainforests or notophyll vine forests; 80–165 m.

Conservation status. The species is considered Endangered, as defined by the Queensland *Nature Conservation Act 1992* (Henderson 1997).

Vernacular name. Sweet myrtle.

ADDITIONAL SPECIMENS EXAMINED. **Australia.** QUEENSLAND. MORETON DISTRICT: Currumbin Ck, 28°S, 153°E, Sep 1980, *Bird s.n.* (BRI AQ344965); Trees Rd and Ducats Rd, Tallebudgera Valley, 28°S, 153°E, Jul 1996, *Booth s.n.* (BRI AQ587553); Brocks Rd near Nicholls Scrub, Currumbin Valley, 28°11′24″S, 153°25′38″E, *Grimshaw G161* (BRI); Currumbin Valley, rainforest, 28°S, 153°E, Oct 1980, *Jones s.n.* (BRI AQ344503, MO 4353023); Currumbin Valley, Nicholl's property, 28°S, 153°E, Mar 1981, *Jones s.n.* (BRI AQ345853); Nicholl's Rainforest, Currumbin Valley, 28°S, 153°E, Nov 1982, *Jones s.n.* (BRI AQ393596); Nichol's Scrub, top end of Brock's Rd off Tomewin Rd, 28°11′S, 153°25′E, *Thomas RFR40 & Barry* (BRI).—NEW SOUTH WALES: Boat Harbour Nature Reserve, NE of Lismore, 28°46′S, 153°20′E, *Bean 7971* (BRI); Lions Park near Lismore, 28°S, 153°E, *Byrnes 3949* (BRI); Boat Harbour, Lismore, 28°S, 153°E, *Byrnes 4057* (BRI); Lions Park near Lismore, 28°S, 153°E, *Byrnes 4067* (BRI); Richmond R., 1876, *Fawcett s.n.* (MEL 726903, 726904); beside Wilson Ck, near Boat Harbour, near Lismore, 28°S, 153°E, *Floyd AGF1630* (BRI, GREE); Boat Harbour in picnic area, *Floyd AGF1030* (BRI, GREE); The Channon, 28°41′S, 153°16′E, *Forster PIF16119 & Tucker* (BRI); Lions Park, N of Lismore, 28°40′S, 153°15′E, *Guymer 2121* (BRI); Wilson Park, Lismore, 28°49′S, 153°16′E, *Harden 81298 & Williams* (BRI, MO); Ballina to Bangalow, Nov 1903, *Maiden & Boorman s.n.* (NSW 144706); Lions Park, ca. 5 km N of Lismore, 28°S, 153°E, Dec 1980, *Williams s.n.* (BRI 345006, CANB, NSW 180736).

CULTIVATED: at 609 Grandview Rd, property of G. Nosworthy, Pullenvale, ca. 3.5 km from turnoff on Moggill Rd, 27°32′S, 152°50′E, *Snow 7749* (BRI, GREE).

The puberulous branchlets and the silvery green leaves, which are capped by a darkish mucro, distinguish *G. fragrantissima*, as do the sweetly fragrant, 4-merous flowers and the red fruit. The oil glands on the adaxial leaf surface are generally faint but visible to the naked eye. The embryo appears to be somewhat oily, but this observation must be confirmed by examining more fresh material.

The label of the holotype as well as of the isotype, in Mueller's herbarium, lists two localities: Moreton Bay in Queensland and Richmond River in New South Wales near the Queensland border.

The specific epithet refers to the strong fragrance produced by the flowers.

23. Gossia sankowskiorum N. Snow & Guymer, sp. nov.—TYPE: AUSTRALIA. Queensland: Cook District, Byerstown Range, S of Lakeland Downs, 15°59′S, 144°49′E, 20 Jan 1986, *G. P. Guymer 2037* (holotype: BRI-2 sheets!; isotypes: BRI! CANB! GREE! K! MO! NSW! QRS!).

Arbores multicaules vel frutices; ramuli rotundati; folia paginis implexis apice retuso vel mucrone brevi lato praedito. Flores tetrameri.

Multi-stemmed shrubs to trees, 3–10 m tall. Bark of trunk smooth, somewhat fluted or fissured, gray and often mottled or with pinkish, reddish, or cream-colored patches. Branchlets rounded to laterally compressed, not winged or youngest branchlets rarely with distinctly winged edges, light brown to light gray, bark smooth and becoming flaky, glabrous, oil glands few to common, indistinct. Leaves slightly discolorous, matte or rarely glossy above, matte below. Stipules type B or type C (see Morphology). Petioles 2–4.5 mm long, light brown, channeled (sometimes deeply so), eglandular or somewhat glandular. Leaf blades 1.5–5.0 (–6.0) cm long, 0.8–2.5 (–3.0) cm wide, narrowly elliptic to ovate, base cuneate, apex obtuse to acute, the tip rounded, sometimes retuse or mucronate, rarely falcate, margins flat; adaxial surface glabrous, oil glands common, prominent only when observed with magnification; midvein flush; abaxial surface glabrous, oil glands common, visible with magnification; secondary and intramarginal veins not visible to the naked eye to prominent, the inner intramarginal vein 1–4 mm from edge. Inflorescence terminal or axillary, solitary; flowers in panicles; anthopodia and metaxyphylls absent. Peduncles 10–15 mm long, rigid, glabrous. Bracteoles 2, 0.7–1 mm long, 0.3–0.6 mm wide, linear to narrowly triangular, rigid, scalelike without midrib, the tips shorter than base of calyx lobes, glabrous to sparsely sericeous, persistent in fruit. Hypanthium 1.5–2.3 mm long, obconic (the base often gradually tapered), glabrous, oil glands abundant. Calyx lobes 4, 1.1–2.2 mm long, triangular to rounded, apex obtuse, greenish; adaxial surface sparsely sericeous to sericeous; abaxial surface glabrous. Petals 4, 2–6 mm long, 2.5–5.5 mm wide, elliptic to widely ovate or obovate, white to yellowish white; adaxial surface glabrous to glabrescent but sparsely ciliate on margins (especially near apex); abaxial surface glabrous, oil glands few to common but often faint. Stamens 75–110, included, filaments 3.5–4.5 mm long; staminal disk glabrescent; anthers 0.5–0.8 mm long, subglobose to globose, the tip of connective with a single gland (often red and prominent). Style 3–5 mm long, glabrous to sparsely sericeous; ovary apex slightly hairy; ovules 10–12 per placenta, placed irregularly. Berry 7–8 mm long, 7–8.5 mm in diameter, subglobose to subcylindrical, base rounded to tapered, glabrous, dark purple to black. Seeds in mature fruit 1–2; rounded, somewhat flattened, smooth, dark brown to dark purple or sometimes mottled with white, micropylar operculum absent, seed coat soft, leathery and thin. Embryo circinate, starchy, oil glands present, endosperm present (clear, colorless, gelatinous); hypocotyl relatively thin, tip near radicle swollen but thickest in

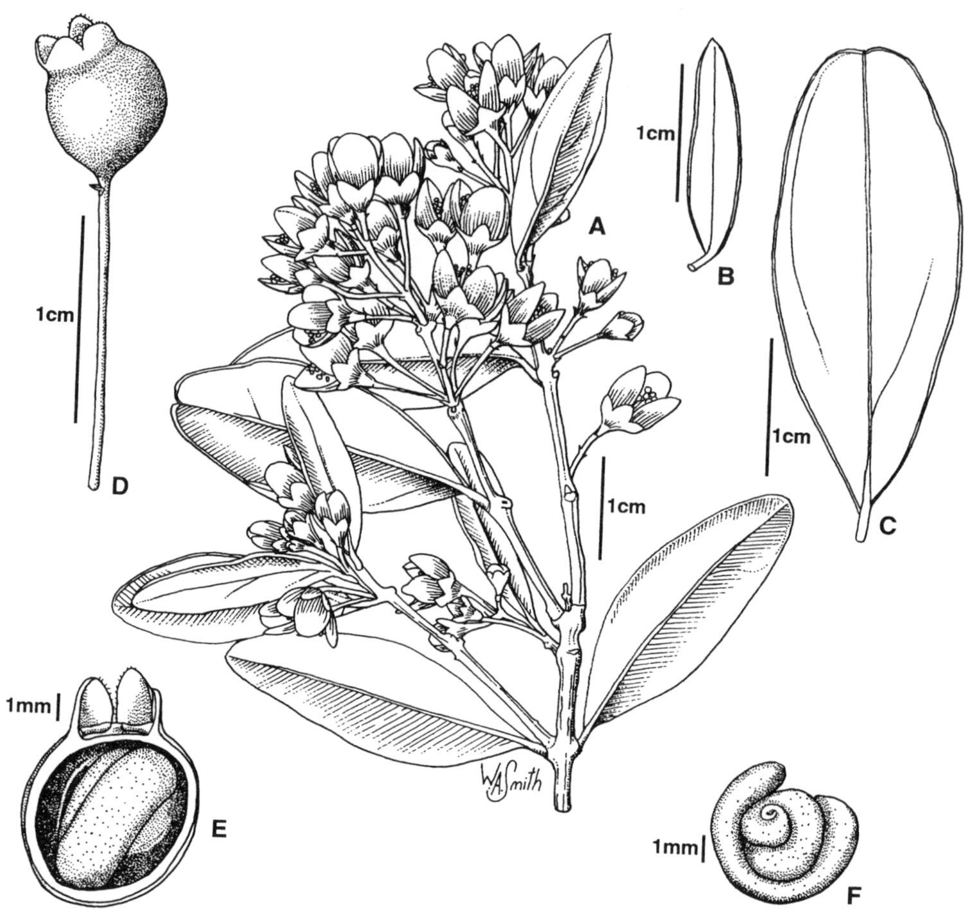

FIG. 23. *Gossia sankowskiorum*. A. Flowering branch. B, C. Variation in leaf shape. D. Fruit and peduncle. E. Fruit in longitudinal section, with lateral view of coiled embryo. F. Embryo. (Based on: A–D, *Sankowsky 1305*; E, F, *Guymer 2036*.)

middle, tip at same plane as embryo; cotyledons mostly straight, shorter than hypocotyl, folded back towards hypocotyl or forward. Fig. 23; Hyland et al. 1999 (see photo).

Phenology. Flowering in November and December; fruiting December through February.

Distribution (Fig. 13). Northeast Queensland, from Coen to Mount Molloy; locally common in deciduous vine thickets or monsoon- and rainforests, mostly in rocky areas; 60–700 m.

Conservation status. The species does not appear to be threatened, but more work is needed to assess overall range and population sizes.

ADDITIONAL SPECIMENS EXAMINED. **Australia.** QUEENSLAND. COOK DISTRICT: Mt Molloy, *Carr 52* (BRI); 11.8 km N of the Palmer R. on the Peninsula Development Rd, 16°01′S, 144°49′E, *Clarkson 4219* (BRI, DNA, K, L, MO, PERTH, QRS); 12 km from the East Normanby river crossing on the Lakeland Downs to Cooktown Rd, 15°41′S, 145°04′S, *Clarkson 6775 & McDonald* (BRI); Bakers Blue Mt, Mona Ck, Font Hills Stn, 16°43′S, 145°10′E, *Fell DF1637* (BRI); Cape Melville NP, 88.9 km NNE of Lakefield HS, 14°11′S,

144°32′E, *Fell DGF3019 & Stanton* (BRI, CANB, DNA, MEL, NSW); Cape Melville NP, Altanmoui Ra. Section, 1.6 km E of Flat Hill, 62.6 km NE of Lakefiled HS, 14°30′S, 144°35′E, *Fell DGF3219 & Stanton* (BRI); Upper Howick R., 48.9 km ENE of Lakefield HS, Kalpowar Pastoral Holding, FR Site 50, Cape York Peninsula, 14°42′S, 144°35′E, *Fell et al. DGF3256* (BRI); 2 km E of Mt Gibson, 16 km SSE of Lakeland Downs, West Normanby R., Cape York Peninsula, 15°59′S, 144°45′E, *Fell DGF3277 & Daunt* (BRI); Birthday Mt, Rokeby NP 23 km N of Coen aerodome, 13°32′S, 143°07′E, *Fell DGF3518 & Jensen* (BRI); Bob's Lookout, Peninsula Development Rd, 16°28′S, 144°53′E, *Ford 1827* (BRI); SFR 143, on ridge above Mcleod R. Falls, 16°27′S, 145°08′E, *Ford 1868* (QRS); Mt Scatterbrain, 5 km E of Lakeland Downs, 15°52′S, 144°54′E, *Forster PIF5563 & Tucker* (BRI); Maitland Downs, 9.5 km S of Laura R. crossing S of Lakeland Downs,16°01′S, 144°49′E, *Forster et al. PIF13432* (BRI); Bakers Blue Mt, Font Hills, 16°41′S, 145°11′E, , *Forster et al. PIF19250* (BRI), *Gray 1596* (BRI, CANB), *Gray 6337* (BRI); Byerstown Ra., *Gray 6530* (BRI); 11.8 km N of Palmer R., Cooktown Rd, 16°00′S, 144°49′E, *Guymer 2036* (BRI, DNA, GREE, MEL, PERTH, UPS); Byerstown Ra., 16°00′S, 144°50′E, *Hyland 8135* (BRI); Maitland Downs Holding, Parish of Byerstown, 16°02′S, 144°49′E, *Hyland 14686* (BRI); ca. 10 km N or the Palmer R. Roadhouse, 16°02′S, 144°49′E, *Hyland 15531* (QRS); Font Hills, 16°43′S, 145°10′E, *Jensen 799* (QRS); Tolga ex Lakeland Downs, 15°S, 144°E, *Sankowsky 1304 & Sankowsky* (BRI); Tolga ex Font Hills, 16°S, 145°E, *Sankowsky 1305, 1306 & Sankowsky* (BRI); Tolga, *Sankowsky 1307 & Sankowsky* (BRI); Tolga ex Lakeland Downs, 15°S, 144°5′E, *Sankowsky 1377 & Sankowsky* (BRI); Tolga ex Font Hills, 16°S, 145°E, *Sankowsky 1383 & Sankowsky* (BRI); between Byerstown Ra. and Lakeland Downs, E of Marreeba-Cooktown Rd, 16°00′S, 144°51′E, *Tracey 14182* (QRS); Bakers Blue Mt, 16°42′S, 145°10′E, *Webb & Tracey 11044* (QRS).

CULTIVATED: Tolga ex Bakers Blue Mountains, Font Hills, *Ford 2542* (GREE); Tolga, ex Lakeland Downs [ca 15°50′S, 144°50′S], *Sankowsky 1241 & Sankowsky* (BRI).

Gossia sankowskiorum is distinguished by the combination of 4-merous flowers, an obconic hypanthium with a sharply tapering base, and the retuse leaf apex (sometimes with a very short mucro). It varies considerably in leaf size, shape, and prominence of the veins, as well as in the color and texture of the bark. Some specimens bear small warts on the leaves, which probably represent a fungal or viral infection (*Tracey 14182*, QRS; *Webb & Tracey 11044*, QRS; *Fell 3256*, BRI). Leaves are occasionally glossy adaxially (e.g., *Fell 3219 & Stanton*). The flower is reported to have a pleasant, sweet aroma (*Ford 2542*, GREE). The apparent presence of gelatinous endosperm should be confirmed by examining additional fresh material.

The specific epithet honors Garry Sankowsky (b. 1943) and Nada Sankowsky (b. 1944), who have made many important collections of this and other species of *Gossia*.

24. Gossia retusa N. Snow & Guymer, sp. nov.—TYPE: AUSTRALIA. Queensland: Cook District, Mapoon, S of Port Musgrave, 11°59′S, 142°52′E, 27 Dec 1980, *A. Morton AM-1007* (holotype: BRI!; isotype: QRS!).

Ramuli quadrangulares; folia pagina superiore nitida nervo medio elevato, apice retuso. Flores tetrameri.

Gnarled shrubs to trees, 1–7 m tall. Bark of trunk mostly smooth or somewhat flaky, whitish or sometimes orangish. Branchlets distinctly 4-angled, edges distinctly winged, light brown to somewhat orangish, smooth but becoming somewhat flaky, glabrous, sometimes with dark, wartlike structures (<1 mm in diameter), oil glands few, indistinct. Leaves discolorous, glossy above, matte below. Stipules type C (see Morphology), the leaf axils sometimes producing a resinous exudate. Petioles 1–3.5 mm long, olive green to brown, slightly channeled, eglandular. Leaf blades 1.5–3.0 (–5.0) cm long, 1.0–2.0 (–3.5) cm wide, elliptic to broadly elliptic or rarely obovate to circular, base cuneate to rounded, apex obtuse, the tip retuse or rarely mucronate, margins revolute; adaxial surface glabrous, oil glands common, usually prominent; midvein raised (extending into petiole); abaxial surface glabrous, oil glands common, often prominent or visible only with magnification; secondary and intramarginal veins not visible to the naked eye or indistinct (venation more prominent above), the inner intramarginal vein 0.5–1.5 mm from edge. Inflorescence terminal or axillary, solitary; flowers solitary or in panicles; anthopodia and metaxyphylls absent. Peduncle 7–12 mm long, rigid, glabrous. Bracteoles 2, ca. 0.5 mm long, ca. 0.2 mm wide, linear to triangular, rigid, scalelike without midrib, the tips not exceeding base of calyx lobes, glabrous, persistent or caducous in fruit. Hypanthium ca. 2 mm long, obconic, glabrous, oil glands abundant. Calyx lobes (3–) 4, 1.5–2.5 mm long, triangular to broadly ovate, apex obtuse, greenish; adaxial surface glabrous; abaxial surface sparsely sericeous. Petals (3–) 4, 3–5 mm long, ca. 3 mm wide, obovate to widely obovate, white; adaxial surface sparsely sericeous on edges and near apex; abaxial surface glabrous, oil glands common to abundant. Stamens ca. 50, included, filaments 2.5–4.0 mm long; staminal disk glabrescent; anthers ca. 0.5 mm long, subglobose to globose, the tip of connective with a single gland. Style 3–3.5 mm long, sparsely sericeous near base; ovary apex slightly hairy; ovules up to 8 per placenta, around edge in single row. Berry 3.5–7 mm long, 3–7 mm in diameter, globose, base rounded, glabrous, dark purple to black. Seeds in mature fruit 1–2, roundish, irregularly indented and rough, purplish or dark brown mottled with some white, micropylar operculum absent, seed coat leathery. Embryo circinate, starchy, oil glands present, endosperm present (especially around cotyledons); hypocotyl relatively thin, tip near radicle not or barely swollen, tip somewhat spiraling above or below plane of embryo; cotyledons straight or folded into U shape, shorter than hypocotyl (very short), pointing forward. Fig. 24.

Phenology. Flowering in November and December; fruiting verified only for December.

Distribution (Fig. 14). Northeastern Queensland; sandy soils in vine thickets or scrub above beachline on stabilized dunes or near swamps and mesic woodlands; 3–70 m.

Conservation status. *Gossia retusa* should be considered Vulnerable until its range and abundance are better known.

ADDITIONAL SPECIMENS EXAMINED. **Australia.** QUEENSLAND. COOK DISTRICT: Horn Is., Torres Strait, 10°35′S, 142°15′E, *Cameron 2220* (QRS); Cape [illegible], NE Australia, *Curdie s.n.* (MEL 60227); 6 km W of the Rocky R. mouth, 36.2 km ENE of Coen, Silver Plains Holding, 13°46′S, 143°29′E, *Fell et al. DGF3485* (BRI); Pascoe R. mouth, north bank, 33.8 km NNW of Lockhart R., Bromley Holding, NE Cape York Peninsula, 12°29′27″S, 143°16′01″E, *Fell DGF3155 & Butcher* (BRI, CANB, DNA); Silver Plains S of Scrubby Ck and W of Colmer Point, 13°46′32″S, 143°29′56″E, *Forster PIF17059* (BRI); Temple Bay Bolt Head, 12°14′59″S, 143°05′02″E, *Forster PIF19364* (BRI); Bolt Head, 12°15′S, 143°05′E, *Gray* 6847 (QRS); Olive R., 12°10′S, 143°05″E, *Hyland 7415*, *7481* (BRI); between Temple Bay and Bolt Head, 12°18′S, 143°05′E, *Hyland 8989* (QRS); 2 km S of Temple Bay Outstation, 12°22′S, 143°05′E, *Hyland 9010* (QRS); Bolt Head, 12°15′S, 143°5′E, *Hyland RFK25985* (BRI); Cape Sidmouth, Eclipse Expedition of 1871, *Moore s.n.* (K, MEL 60220); between Pine R. Basin and Gulf Coast, N of Duyfken Pt., 12°27′S, 141°37′E, *Morton 1125* (QRS); Nesbit R., 13°26′S, 143°10′E, *Tracey 14229* (BRI); N of Oliver R., 12°07′S, 143°07′E, *Webb & Tracey Pen10A* (QRS); half mile N of Olive R., 12°20′S, 143′05′E, *Webb & Tracey Pen11B* (QRS); 5 mi N of crossing on Massey Ck on rd between Silver Plains Stn and Rocky R., 13°50′S, 143°29′E, *Webb & Tracey 9717A* (BRI); Olive R., 12°07′S, 143°05′S, *Webb & Tracey 11012* (QRS); Starke Stn HS, Cape Flattery Rd, near beach, ca. 15°00′S, 145°10′E, *Webb & Tracey 13537* (BRI); ca. 1 km up river from Pascoe R. mouth, 12°30′S, 145°15′E, *Webb & Tracey 13538* (BRI).—NORTH KENNEDY DISTRICT: NPR 95 Hinchinbrook Is., Shepherd Bay, 18°15′S, 146°15′E, *Hyland 10615* (QRS).

CULTIVATED: Tolga ex Bolt Head, 17.13°S, 145.28°S, *Ford 2124* (GREE).

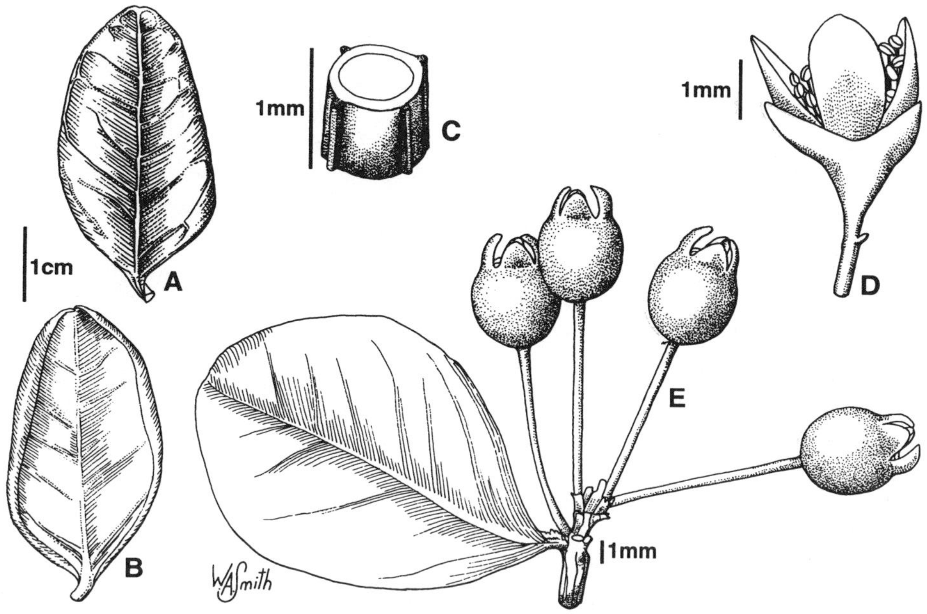

FIG. 24. *Gossia retusa*. A. Leaf, adaxial view. B. Leaf, abaxial view, with recurved margins. C. Stem cross section showing wings. D. Opening flower. E. Fruiting branchlet. (Based on: A-C, *Hyland 7415*; D, *Moore s.n., MEL 60220*; E. *Morton AM-1007*.)

Young growth and leaves of *G. retusa* are a light reddish brown. The species most resembles *G. gonoclada*, but differs in its 4-merous flowers, glabrous peduncles and hypanthium, narrower bracteoles, distinctly quadrangular stem (i.e., not merely winged), and shorter, retuse, distinctly revolute leaves. The two species are entirely disjunct.

Flowering material is known only from *MEL 60220* (collector unknown) and *Ford 2124* (QRS, BRI), and fruiting material only from the type and *MEL 60220*. Andrew Ford (pers. comm.) reports that the styles are reddish or pink, and that petals can lack oil glands altogether. Some individuals appear to produce a resinous exudate in the axils of young leaves (*Tracey 14229*, BRI; *Hyland 7481*, BRI), which, in Myrtaceae, van Steenis (1968: 358) and Weberling (1985) believed is derived from the stipules.

The epithet *retusa* refers to the rounded and notched apex of the leaf.

25. Gossia byrnesii N. Snow & Guymer, sp. nov.—TYPE: AUSTRALIA. Queensland: Cook District, cultivated plant, originally collected from Isabella Falls, Rainforest type 4b, 22 Dec 1988, *G. Sankowsky 959* & *N. Sankowsky* (holotype: BRI!).

Ramuli quadrangulares; folia pagina superiore nitida nervo medio non elevato, apice retuso vel vix mucronato. Flores tetrameri.

Trees, height unknown. Branchlets rounded or slightly 4-angled, not winged, brown, smooth, glabrous to very sparsely sericeous, oil glands indistinct, common. Leaves concolorous, slightly glossy above, matte below. Stipules type B (see Morphology). Petioles 1–1.5 mm long, orangish to brown, channeled slightly or not, glandular

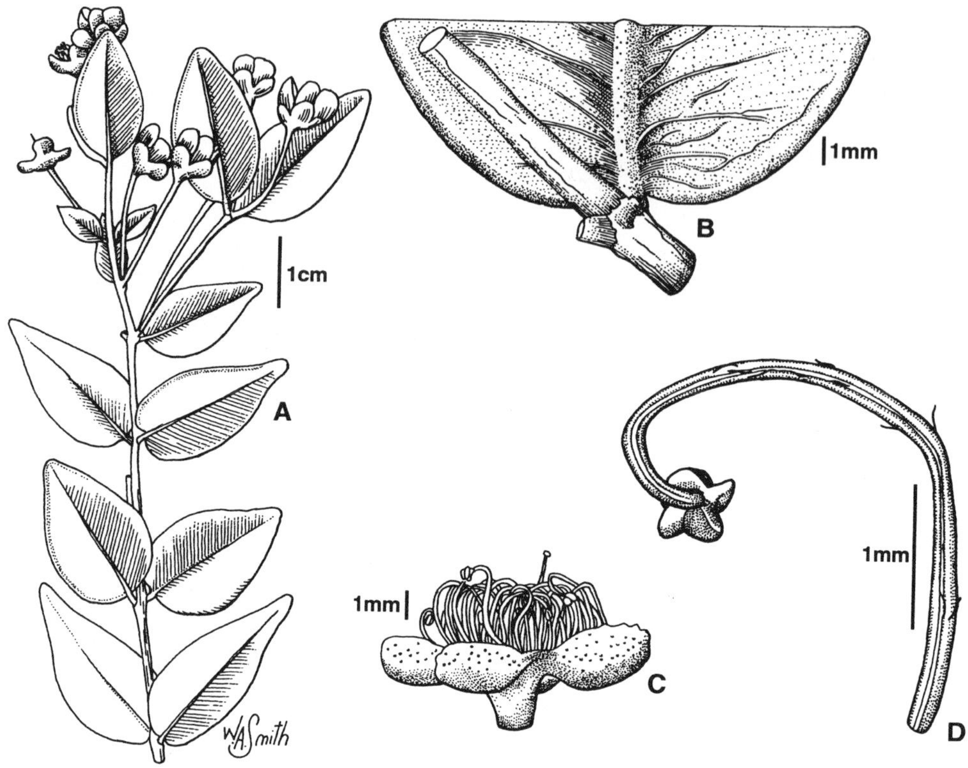

FIG. 25. *Gossia byrnesii*. A. Flowering branch. B. Proximal part of leaf blade, adaxial view, and portion of branchlet. C. Flower. D. Filament bearing anther and scattered hairs. (Based on *Sankowsky 959*.)

when young. Leaf blades 1.7–3.2 cm long, 0.9–1.6 cm wide, narrowly ovate, base rounded to slightly cordate, apex obtuse to acute, the tip rounded to pointed, margins flat; adaxial surface glabrous, oil glands abundant, visible with magnification; midvein flush; abaxial surface glabrous, oil glands few, visible with magnification; secondary veins indistinct; intramarginal veins surface not visible with the naked eye or indistinct, the inner intramarginal vein 0.5–1 mm from edge. Inflorescence terminal or axillary, a monad or consisting of 2 flowers on glabrous peduncles; anthopodia and metaxyphylls absent. Peduncles 12–14 mm long, rigid, glabrous. Bracteoles 2, 1–1.5 mm long, ca. 0.5 mm wide, linear to narrowly triangular, rigid, scalelike without midrib, the tips shorter than base of calyx lobes, glabrous; persistence in fruit unknown. Hypanthium 2–3 mm long, obconic to cupulate, glabrous, oil glands common. Calyx lobes 4, 1.5–3 mm long but sometimes very unequal, broadly ovate to rounded, apex obtuse, greenish; adaxial surface sparsely villous; abaxial surface glabrous. Petals 4, 3–4.2 mm long, 2.7–3 mm wide, widely elliptic to obovate, white; adaxial surface glabrous; abaxial surface villous, oil glands common but faint. Stamens 60–80, included, filaments ca. 3 mm long (and sparsely sericeous); staminal disk sparsely shortly hairy; anthers ca. 0.5 mm long, subglobose to globose, the tip of connective eglandular. Style 3–3.5 mm long, glabrous; ovary apex slightly hairy; ovules 5–9 per placenta, placed irregularly. Fruit unknown. Fig. 25.

Conservation status. The species is considered Critically Endangered (criteria B and

D, Species Survival Commission 1994).

Gossia byrnesii is known only from the type specimen, which was cultivated from material collected at Isabella Falls (Fig. 18) in rainforests at an elevation of 260 m. We place this new species in *Gossia*, but examination of fruiting material is necessary to confirm this disposition.

Gossia byrnesii is distinctive in that the branchlets are rounded to slightly quadrangular, the 4-merous sepal lobes are often quite unequal, and the filaments are often sparsely sericeous. The adaxial leaf surface is glossy with a flush midvein, and has small, abundant oil glands.

The specific epithet honors Norman Byrnes (1922–1998), who made important collections of *Gossia* and *Lenwebbia*, including the type specimen of *L. prominens*.

Doubtful and Excluded Names

The names here referred to *Gossia* apply to species from New Caledonia.

Austromyrtus alaternoides (Brongniart & Gris) Burret, Notizbl. Bot. Gart. Berlin-Dahlem 15: 504. 1941. = *Gossia* sp.

Austromyrtus aneityensis (Guillaumin) Burret, Notizbl. Bot. Gart. Berlin-Dahlem 15: 506. 1941. = *Gossia* sp.

Austromyrtus aphthosa (Vieillard ex Brongniart & Gris) Burret, Notizbl. Bot. Gart. Berlin-Dahlem 15: 504. 1941. = *Gossia* sp.

Austromyrtus cataractarum (Guillaumin) Burret, Notizbl. Bot. Gart. Berlin-Dahlem 15: 503. 1941. = *Gossia* sp.

Austromyrtus clusioides (Brongniart & Gris) Burret, Notizbl. Bot. Gart. Berlin-Dahlem 15: 503. 1941. = *Gossia* sp.

Austromyrtus conspicua (Vieillard ex Guillaumin) Burret, Notizbl. Bot. Gart. Berlin-Dahlem 15: 505. 1941. = *Gossia* sp.

Austromyrtus diversifolia (Brongniart & Gris) Burret, Notizbl. Bot. Gart. Berlin-Dahlem 15: 503. 1941. = *Gossia* sp.

Austromyrtus exaltata (F. M. Bailey) Burret, Notizbl. Bot. Gart. Berlin-Dahlem 15: 501. 1941. = *Syzygium luehmannii* (F. M. Bailey) L. Johnson.

Austromyrtus gatopensis (Guillaumin) Burret, Notizbl. Bot. Gart. Berlin-Dahlem 15: 505. 1941. = *Eugenia gatopensis* Guillaumin, Bull. Soc. Bot. France 85: 638. 1938.

Austromyrtus gomonenensis (Guillaumin) Burret, Notizbl. Bot. Gart. Berlin-Dahlem 15: 501. 1941, pro syn.

Austromyrtus gyrosepala (E. G. Baker) Burret, Notizbl. Bot. Gart. Berlin-Dahlem 15: 506. 1941. = *Eugenia gyrosepala* E. G. Baker, J. Linn. Soc. 45: 313. 1921.

Austromyrtus horizontalis (Pancher ex Brongniart & Gris) Burret, Notizbl. Bot. Gart. Berlin-Dahlem 15: 504. 1941. = *Eugenia horizontalis* Pancher ex Brongniart ex Gris.

Austromyrtus kanalaensis (Hoch.) Burret, Notizbl. Bot. Gart. Berlin-Dahlem 15: 504. 1941. = *Uromyrtus* sp.

Austromyrtus kuakuensis (E. G. Baker) Burret, Notizbl. Bot. Gart. Berlin-Dahlem 15: 503. 1941, pro syn.

Austromyrtus lotoides (Vieillard ex Guillaumin) Burret, Notizbl. Bot. Gart. Berlin-Dahlem 15: 504. 1941. = *Gossia* sp.

Austromyrtus luteo-viridis (E. G. Baker) Burret, Notizbl. Bot. Gart. Berlin-Dahlem 15: 505. 1941. = *Gossia* sp.

Austromyrtus mendute (Guillaumin) Burret, Notizbl. Bot. Gart. Berlin-Dahlem 15: 505. 1941. = *Uromyrtus* sp.

Austromyrtus metrosideros (F. M. Bailey) Burret, Notizbl. Bot. Gart. Berlin-Dahlem 15: 501, 1941. = *Uromyrtus metrosideros* (F. M. Bailey) A. J. Scott.

Austromyrtus nigripes (Guillaumin) Burret, Notizbl. Bot. Gart. Berlin-Dahlem 15: 505. 1941. = *Gossia* sp.

Austromyrtus pancheri (Brongniart & Gris) Burret, Notizbl. Bot. Gart. Berlin-Dahlem 15: 503. 1941. = *Gossia* sp.

Austromyrtus ploumensis (Daniker) Burret, Notizbl. Bot. Gart. Berlin-Dahlem 15: 504. 1941. = *Gossia* sp.

Austromyrtus poimbailensis (Guillaumin) Burret, Notizbl. Bot. Gart. Berlin-Dahlem 15: 505. 1941. = *Eugenia* sp.

Austromyrtus prolixa (E. G. Baker) Burret, Notizbl. Bot. Gart. Berlin-Dahlem 15: 505. 1941. = *Uromyrtus* sp.

Austromyrtus stricta (Pancher ex Brongniart & Gris) Burret, Notizbl. Bot. Gart. Berlin-Dahlem 15: 505. 1941. = *Uromyrtus* sp.

Austromyrtus styphelioides (Schlechter) Burret, Notizbl. Bot. Gart. Berlin-Dahlem 15: 504. 1941. = *Gossia* sp.

Austromyrtus vieillardi (Brongniart & Gris) Burret, Notizbl. Bot. Gart. Berlin-Dahlem 15: 503. 1941. = *Gossia* sp.

ACKNOWLEDGMENTS

We thank the curators of A, ASU, BISH, BRI, CANB, CONN, K, LE, MEL, MO, NSW, NY, QRS, SING, SP, RSA, and US for loans and permission to examine specimens. We are grateful to A. Ford for sharing numerous personal observations of living material; W. Smith for illustrations, L. Pedley and W. R. Anderson for assistance with Latin; C. A. Jarvis (BM), J. Grimes (MEL), E. NicLughada (K), and E. Woods (A) for confirming or checking type specimens; A. R. Bean (BRI) for assisting with older literature from Queensland; P. Sharpe and M. Quirk for translations from German; P. Bostock (BRI) for generating the high quality maps; L. Craven for assistance at CANB; P. Wilson for collecting seed, discussions on generic boundaries and phylogenetic relationships, and checking a type specimen; L. Landrum for discussions on generic boundaries and character delimitation; J. Dawson for discussion of species boundaries in New Caledonian members of *Gossia*, B. Simon for assistance with DELTA; P. Forster, R. Fensham, and W. J. F. McDonald for discussions regarding variation in *G. bidwillii;* H. Winters (BRI), M. Stiffler (MO), and the library staff at the University of Northern Colorado; P. Cameron for permission to collect specimens at the Brisbane Botanic Gardens Mt Coot-tha; G. Nosworthy, G. Dorrington (Hyatt Regency in Coolum Beach), and J. P. Raibier for permission to collect on their properties; J. Bradford, S. Snow, S. Philips, P. Sharpe, and H. Dillewaard for assistance in the field. We also thank L. Landrum, A. Ford, and C. Anderson for their constructive suggestions for improving the manuscript. Research was supported by grants from the Australian Biological Resources Study during 1997–1999. The University of Northern Colorado granted the senior author a leave of absence in 1999 to work on this project.

LITERATURE CITED

Alberch, P., S. J. Gould, G. F. Oster, and D. W. Wake. 1979. Size and shape in ontogeny and phylogeny. Paleobiology 5: 296–317.

Australian Biological Resources Study. 1993. *Flora of Australia: guide for contributors*. Canberra: ABRS.

Bailey, F. M. 1900. *The Queensland Flora. Part II. Connaraceae to Cornaceae*. Brisbane: Queensland Government.

Baum, D. A., and M. J. Donoghue. 1995. Choosing among alternative "phylogenetic" species concepts. Syst. Bot. 20: 560–573.

Bean, A. R. 1999. A revision of the *Babingtonia virgata* (J. R. Forst. & G. Forst.) F. Mueller complex (Myrtaceae) in Australia. Austrobaileya 5: 157–171.

Bentham, G. 1867. *Flora Australiensis*, vol. 3. London: Lovell Reeve & Co.

______. 1869. Notes on Myrtaceae. J. Linn. Soc. London, Bot. 10: 101–166.

Berg, O. 1855–1856. Revision Myrtacearum Americae. Linnaea 27: 1–472.

Blake, S. T. 1949. The specimen, the species and the botanist. Austral. J. Sci. 11: 119–123.

Bradford, J. C. 1998. A cladistic analysis of species groups in *Weinmannia* (Cunoniaceae) based on morphology and inflorescence architecture. Annals Missouri Bot. Gard. 85: 565–593.

Briggs, B. A, and L. A. S. Johnson. 1979. Evolution in the Myrtaceae—evidence from inflorescence structure. Proc. Linn. Soc. of N. S. Wales 102: 157–256.

Brophy, J. J., and J. C. Doran. 1996. *Essential oils of tropical Asteromyrtus, Callistemon, and Melaleuca species*. Canberra: Australian Center for International Agricultural Research.

Brophy, J. J., R. J. Goldsack, C. J. R. Fookes, and P. I. Forster. 1995a. The essential oils of Australian *Austromyrtus* sens. lat. Part 1. The *A. dulcis* group. Flav. Fragr. J. 10: 69–73.

Brophy, J. J., R. J. Goldsack, and P. I. Forster. 1995b. The essential oils of Australian *Austromyrtus* sens. lat. Part 2. The *Austromyrtus lasioclada* group. Flav. Fragr. J. 10: 293–296.

______. 1995c. (E)-β-Ocimene from two species of *Austromyrtus* (Myrtaceae). J. Essential Oil Res. 7: 1–4.

Brophy, J. J., R. J. Goldsack, C. J. R. Fookes, and P. I. Forster. 1996a. The essential oils of Australian *Austromyrtus* sens. lat. Part 3. The *Austromyrtus bidwillii* group. Flav. Fragr. J. 11: 275–287.

Brophy, J. J., R. J. Goldsack, and P. I. Forster 1996b. The essential oils of Australian species of *Uromyrtus* (Myrtaceae). Flav. Fragr. J. 11: 133–138.

______. 1996c. Variation in *Archirhodomyrtus beckleri* (F. Mueller) A. J. Scott (Myrtaceae): evidence from volatile oils. Flav. Fragr. J. 11: 11–14.

______. 1997a. The essential oils of Australian species of *Rhodomyrtus* (Myrtaceae). Flav. Fragr. J. 12: 103–108.

______. 1997b. The essential oils of Australian species of *Rhodamnia* (Myrtaceae). Flav. Fragr. J. 12: 345–354.

Burret, M. 1941. Myrtaceen-Studien. Notizbl. Bot. Gart. Berlin-Dahlem 15: 479–550.

Byrnes, N. 1982. The genus *Myrtus* or *Austromyrtus* in Australia? Austral. Soc. Syst. Bot. Newsletter 31: 10–11.

Candolle, A. P. de. 1828. Myrtaceae. In *Prodromus systematis naturalis regni vegetabilis* by A. P. de Candolle, 3: 207–296. Paris: Treuttel et Würtz.

Chappill, J. A., and P. Y. Ladiges. 1996. Phylogenetic analysis of *Eucalyptus* informal subgenus *Symphyomyrtus* section *Maidenaria*. Austral. Syst. Bot. 9: 71–93.

Chippendale, G.M. 1988. *Myrtaceae—Eucalytpus*, *Angophora. Flora of Australia* 19: 1–542. Canberra: Australian Government Publishing Service.

Claridge, M. F., H. A. Dawah, and M. R. Wilson, eds. 1997. *Species: the units of biodiversity*. London; New York: Chapman & Hall.

CPITT (Center for Pest Information Technology and Transfer). 1999. *Lucid Professional: contemporary identification tools*. Melbourne: CSIRO Publishing.

Cracraft, J. 1983. Species concepts and species analysis. In *Current ornithology*, ed. R. Johnson, 159–187. New York: Plenum Press.

———. 1992. The species of the birds-of-paradise (Paradisaeidae): Applying the phylogenetic species concept to a complex pattern of diversification. Cladistics 8: 1–43.

———. 2000. Species concepts in theoretical and applied biology: A systematic debate with consequences. In *Species concepts and phylogenetic theory: a debate*, ed. Q. D. Wheeler and R. Meier, 3–14. New York: Columbia University Press.

Crisp, M. D., H. P. Linder, and P. H. Weston. 1995. Cladistic biogeography of plants in Australia and New Guinea: Congruent patterns reveals two endemic tropical tracks. Syst. Biol. 44: 457–473.

Crisp, M. D., J. G. West, and H. P. Linder. 1999. Biogeography of the terrestrial flora. *Flora of Australia*, 2d ed., 1: 321–367. Canberra: Australian Biological Resources Study/CSIRO.

Dallwitz, M., T. A. Paine, and E. J. Zurcher. 1993. *User's guide to the DELTA system: a general system for processing taxonomic descriptions*, 4th ed. Canberra: CSIRO Division of Entomology.

Davis, J. I. 1997. Evolution, evidence, and the role of species concepts in phylogenetics. Syst. Bot. 22: 373–403.

———. 1999. Monophyly, populations and species. In *Molecular systematics and plant evolution*, ed. P. M. Hollingsworth, R. M. Bateman, and R. J. Gornall, 139–170. [Systematics Association Special Volume 57] London: Francis and Taylor.

Davis, J. I., and P. S. Manos. 1991. Isozyme variation and species delimitation in the *Puccinellia nuttalliana* complex (Poaceae): an application of the phylogenetic species concept. Syst. Bot. 16: 431–445.

Davis, J. I., and K. C. Nixon. 1992. Populations, genetic variation, and the delimitation of phylogenetic species. Syst. Biol. 41: 421–435.

de Pinna, M. C. C. 2000 ["1999"]. Species concepts and phylogenies. Rev. Fish Biol. Fisheries 9: 353–373.

de Queiroz, K. 1998. The general lineage concept of species, species criteria, and the process of speciation: A conceptual unification and terminological recommendations. In *Endless forms: species and speciation*, ed. D. J. Howard and S. H. Berlocher, 57–75. New York: Oxford University Press.

———. 1999. The general lineage concept of species and the defining properties of the species category. In *Species: new interdisciplinary essays*, ed. R. A. Wilson, 49–89. Cambridge, Mass.: MIT Press.

Domin, K. 1928. *Myrtus* L. Biblioth. Bot. 89: 1028–1029.

Doyle, J. J. 1992. Gene trees and species trees: Molecular systematics as one-character taxonomy. Syst. Bot. 17: 144–163.

Ereshefsky, M., ed. 1992. The units of evolution: essays on the nature of species. Cambridge, Mass.: MIT Press.

Fensham, R. J. 1995a. The floristics and structure of dry rainforest at Forty Mile Scrub National Park, north Queensland. Cunninghamia 4: 483–495.

———. 1995b. Floristics and environmental relations of inland dry rainforest in north Queensland, Australia. J. Biogeogr. 22: 1047–1063.

Flora of Australia. 1999. Vol. 1, Introduction, 2d ed. Melbourne: ABRS/CSIRO.

Floyd, A. G. 1973. New South Wales rainforest trees. Part III, Family Myrtaceae. Coffs Harbour: Forestry Commission of New South Wales, Research Note 28.

———. 1989. Rainforest trees of mainland south-eastern Australia. Sydney: Inkata Press.

Floyd, S. K., and W. E. Friedman. 2000. Evolution of endosperm developmental patterns among basal flowering plants. Int. J. Plant Sci. 161(6 Suppl.): S57–S81.

Forster, P. I., P. D. Bostock, L. H. Bird, and A. R. Bean. 1991. *Vine forest plant atlas for south-east Queensland with assessment of conservation status*. Indooroopilly: Queensland Herbarium.

Frost, D. R., and A. G. Kluge. 1994. A consideration of epistemology in systematic biology, with special reference to species. Cladistics 120: 259–294.

Gentry, A. W. 1993. *A field guide to the families and genera of woody plants of northwest South America (Columbia, Ecuador, Peru) with supplementary notes on herbaceous taxa*. Washington, D.C.: Conservation International.

Gift, N., and P. F. Stevens. 1997. Vagaries in the delimitation of character states in quantitative variation—an experimental study. Syst. Biol. 46: 112–125.

Gould, S. J. 1989. *Wonderful life: the Burgess Shale*. New York: Norton & Sons.

Grifo, F. T. 1992. A revision of *Myrcianthes* Berg (Myrtaceae). Unpublished Doctoral Dissertation, Cornell University, Ithaca, New York.

Guillaumin, A. 1931. Contributions to the flora of the New Hebrides: Plants collected by S. F. Kajewski in 1928 and 1929. J. Arnold Arbor. 12: 221–264.

———. 1938. Matériaux pour la flore de la Nouvelle-Calédonie. LIII. Revision des Myrtacées à fruit charnu précédée de quelques notes supplémentaires sur les Myrtacées à fruit sec (1). Bull. Soc. Bot. France 85: 626–653.

Guymer, G. P. 1988. A new species of *Backhousia* Hook. & Harvey (Myrtaceae) from Queensland with a reappraisal of *Backhousia floribunda* A. J. Scott. Austrobaileya 2: 567–569.

Henderson, R. F. J. 1997. *Queensland plants: names and distribution*. Brisbane: Queensland Department of Environment.

Hewson, H. J. 1988. *Plant indumentum. A handbook of terminology*. Australian Flora and Fauna Series No. 9. Canberra: Australian Government Publishing Service.

Hennig, W. 1966. *Phylogenetic systematics*. Urbana: University of Illinois Press.

Hickey, L. J. 1973. Classification of the architecture of dicotyledonous leaves. Amer. J. Bot. 60: 17–33.

Hooker, J. D. 1864. *Handbook of the New Zealand Flora*. London: Lovell Revee & Co.

Howard, D. J., and S. H. Berlocher. 1998. *Endless forms: species and speciation*. New York: Oxford University Press.

Hull, D. L. 1965. The effect of essentialism on taxonomy—two thousand years of stasis (II). Brit. J. Phil. Sci. 16: 1–18.

Humphries, C. J., and P. M. Richardson. 1980. Hennig's methods in phytochemistry. In *Chemosystematics: principles and practice*, ed. F. A. Bisby, J. G. Vaughan, and C. A. Wright, 353–378. [Systematics Association Special Volume 16] London: Academic Press.

Hyland, B. P. M. 1983. A revision of *Syzygium* and allied genera (Myrtaceae) in Australia. Austral. J. Bot., Suppl. Ser. 9: 1–164.

Hyland, B. P. M., and T. Whiffin. 1993. *Australian tropical rainforest trees: an interactive identification system*, vol. 2. Melbourne: CSIRO Publications.

Hyland, B. P. M., T. Whiffin, D. C. Christophel, B. Gray, R. W. Elick, and A. J. Ford. 1999. *Australian tropical rainforest trees and shrubs*. Collingwood, Austraila. CSIRO Publications.

Johnson, L. A. S., and B. G. Briggs. 1985 ["1984"]. Myrtales and Myrtaceae—a phylogenetic analysis. Annals Missouri Bot. Gard. 71: 700–756.

Jones, D. L. 1986. *Ornamental rainforest plants of Australia*. Frenchs Forest, N. S. W.: Reed Books Pty Ltd.

Kausel, E. 1956. Beitrag zur Systematik der Myrtaceen. Ark. Bot., ser. 2, 3: 491–516.

Kawasaki, M. L. 2000. A new species of *Campomanesia* (Myrtaceae) from Brazil. Brittonia 52: 188–190.

Klucking, E. P. 1988. Myrtaceae. In *Leaf venation patterns* by E. P. Klucking, vol. 3. Berlin: J. Cramer.

Landrum, L. R. 1981. A monograph of the genus *Myrceugenia* (Myrtaceae). Fl. Neotrop. Monogr. 29: 1–135.

———. 1986. *Campomanesia*, *Pimenta*, *Blepharocalyx*, *Legrandia*, *Acca*, *Myrrhinium*, and *Luma* (Myrtaceae). Fl. Neotrop. Monogr. 45: 1–178.

———. 1988. Systematics of *Myrteola* (Myrtaceae). Syst. Bot. 13: 120–132.

———. 1989. The generic position of *Myrtus alternifolia* and notes on *Calycolpus* (Myrtaceae). Madroño 37: 9–14.

Landrum, L. R., and D. Stevenson. 1986. Variability of embryos in subtribe Myrtinae (Myrtaceae). Syst. Bot. 11: 155–162.

Landrum, L. R., and M. L. Kawasaki. 1997. The genera of Myrtaceae in Brazil: an illustrated synoptic treatment and identification keys. Brittonia 49: 508–536.

Legrand, C. D. 1975. Sobre *Pilothecium* (Kiaersk.) Kaus. Bradea 2: 33–40.

Mayr, E. 1982. *The growth of biological thought: divergence, evolution, and inheritance*. Cambridge: Belknap Press of Harvard University.

McVaugh, R. 1956a. Tropical American Myrtaceae: notes on generic concepts and descriptions of previously unrecognized species. Fieldiana, Bot. 29(3): 145–228.

———. 1956b. Nomenclatural notes on Myrtaceae and related families. Taxon 5: 133–147, 162–167.

———. 1963. Tropical American Myrtaceae II: notes on generic concepts and descriptions of previously unrecognized species. Fieldiana, Bot. 29(8): 393–532.

———. 1968. The genera of American Myrtaceae—an interim report. Taxon 17: 354–418.

———. 1969. Myrtaceae. In "The botany of the Guayana Highland—Part VIII" by Bassett Maguire and collaborators. Memoirs N. Y. Bot. Gard. 18(2): 55–286.

Niedenzu, F. 1893. Myrtaceae. In *Die Natürlichen Pflanzenfamilien*, ed. A. Engler and K. Prantl, Pt. 3, 7: 57–105. Leipzig: Wilhelm Engelmann.

Nixon, K. C., and Q. D.Wheeler. 1990. An amplification of the phylogenetic species concept. Cladistics 6: 211–223.

Nixon, K. C., and J. I. Davis. 1991. Polymorphic taxa, missing values and cladistic analysis. Cladistics 9: 413–426.

Patterson, C. 1982. Morphological characters and homology. In *Problems in phylogenetic reconstruction*, ed. K. A. Josey and A. E. Friday, 21–74. [Systematics Association Special Volume 21] London: Academic Press.

Peterson, A. T., and A.G. Navarro-Sigüenza, 1999. Alternate species concepts as bases for determining priority conservation areas. Conserv. Biol. 13: 427–431.

Petit, L.-A. 1908. Recherches sur la structure anatomique du fruit et de la graine des Myrtacées. (Thesis, Université de Paris). Lucien Declume, Lons-le-Saunier.

Salywon, A., N. Snow, and L. R. Landrum. 2002. Phylogenetic relationships of the berry fruited Myrtaceae as inferred from ITS sequences. Abstracts 149. Botany 2002; American Society of Plant Taxonomists.

Schmid, R., and P. Baas. 1984. The occurrence of scalariform perforation plates and helical vessel wall thickenings in wood of Myrtaceae. IAWA Bull., n. s., 5(3): 197–215.

Scott, A. J. 1979. New species and combinations in Myrtaceae from Malesia and Australia. Kew Bull. 33: 511–515.

———. 1980. A new species of *Myrtus* from the Solomon Islands. Kew Bull. 35: 412.

———. 1986. A new combination in *Uromyrtus* (Myrtaceae) from Australia. Kew Bull. 41: 286.

Shapcott, A., and J. Playford. 1996. Comparison of genetic variability in remnant and wide-spread rainforest understorey species of *Austromyrtus* (Myrtaceae). Biodiv. Conserv. 5: 881–895.

Smith, J. E. 1797. Botanical characters of some plants of the natural order of *Myrti*. Trans. Linn. Soc. London 3: 255–288.

Smith, L. S. 1956. New species of and notes on Queensland plants. Proc. Royal Soc. Queensland 67: 29–40.

Snow, N. 1997. Application of the phylogenetic species concept: a botanical monographic perspective. Austrobaileya 5: 1–8.

———. 1998. The use of hairs for phylogenetic inference in grasses (Poaceae). Abstract 51. Monocots II: Second International Conference on the Comparative Biology of the Monocotyledons and Third International Symposium on Grass Systematics and Evolution. Sydney: Royal Botanic Gardens.

———. 1999. Notes on generic concepts in *Rhodomyrtus*, *Archirhodomyrtus*, *Decaspermum*, and *Pilidiostigma* (Myrtaceae). Austral. Syst. Bot. Soc. Newsletter 99: 5–7.

———. 2000. Conspectus of Australasian Myrtinae (Myrtaceae). Kew Bull. 22: 647–654.

Snow, N., and G. P. Guymer.1999. Systematic and cladistic studies of *Myrtella* F. Mueller and *Lithomyrtus* F. Mueller (Myrtaceae). Austrobaileya 5: 173–207.

———. 2001a. Revision of the Australian species of *Uromyrtus* (Myrtaceae) and a new combination for New Caledonia. Syst. Bot. 26: 733–742.

———. 2001b. A new and endangered species of *Decaspermum* (Myrtaceae) from east-central Queensland. Novon 11: 475–478.

Snow, N., and P. L. Keating. 1999. The relevance of specimen citations to conservation. Conserv. Biol. 13: 943–944.

Species Survival Commission. 1994. *IUCN Red List categories*. Gland: IUCN Council.

Stanley, T. D., and E. M. Ross. 1986. *Austromyrtus* (Nied.) Burret. In *Flora of south-eastern Queensland*, ed. T. D. Stanley and E. M. Ross, 2: 207–208. Brisbane: Queensland Department of Primary Industries.

Steenis, C. G. G. J. van 1957. Specific and infraspecific delimitation. In *Flora Malesiana*, ed. C. G. G. J. van Steenis, I, 5(3): clxvii–ccxxxiv. Groningen: P. Noordhoff.

———. 1968. Reduction of the genus *Kania* Schltr. to *Metrosideros* (Myrtaceae). Blumea 16: 357–359.

Stevens, P. F. 1997. J. D. Hooker, George Bentham, Asa Gray, and Ferdinand Mueller on species limits in theory and practice: a mid-Nineteenth Century debate and its repercussions. Hist. Records Austral. Sci. 11: 345–370.

Systematics Association Committee for Descriptive Biological Terminology. 1962. II. Terminology of simple symmetrical plane shapes (chart 1). Taxon 11: 145–156.

Takhtajan, A. 1986. *Floristic regions of the world*. Berkeley: University of California Press.

van Wyk, A. E., and R. Botha, R. 1984. The genus *Eugenia* (Myrtaceae) in southern Africa: ontogeny and taxonomic value of the seed. S. Afr. J. Bot. 3: 63–80.

Wagner, G. P. 1989. The origin of morphological characters and the biological basis of homology. Evolution 43: 1157–1171.

Webb, L. J. 1959. A physiognomic classification of Australian rain forests. J. Ecol. 47: 551–570.

———. 1968. Environmental relationships of the structural types of Australian rainforest. Ecology 49: 296–311.

Webb, L. J., J. G. Tracey, and W. T. Williams. 1984. A floristic framework of Australian rainforests. Austral. J. Ecol. 9: 169–198.

Weberling, F. 1956. Untersuchungen über rudimentäre Stipeln bei den *Myrtales*. Flora 143: 201–218.

———. 1984 ["1985"]. Stipular structures. In "The Order Myrtales: Circumscription, variation, and relationships," ed. R. Dahlgren and R. F. Thorne. Annals Missouri Bot. Gard. 71: 633–699.

———. 2000. *Tepualia stipularis* (Myrtaceae), una planta con apéndices estipuliformes y estípulas rudminentarias. Kurtziana 28: 313–315.

Wheeler, Q. R., and R. Meier. 2000. *Species concepts and phylogenetic theory: a debate*. New York: Columbia University Press.

White, C. T. 1939. Contributions to the Queensland Flora, No. 6. Proc. Royal Soc. Queensland 50: 66–87.

Wiens, J. J., and M. R. Servedio. 2000. Species delimitation in systematics: inferring diagnostic differences between species. Proc. Royal Soc. London (Biology) 267: 633–636.

Williams, K. A. W. 1984. *Native plants of Queensland*, vol. 2. Brisbane: Printercraft.

Wilson, J. 1999. *Biological individuality: the identity and persistence of living entities*. Cambridge; New York: Cambridge University Press.

Wilson, K. A. 1960. The genera of Myrtaceae in the southeastern United States. J. Arnold Arbor. 41: 270–278.

Wilson, P. G. 1991. *Austromyrtus*. In *Flora of New South Wales*, ed. G. J. Harden, 2: 192–194. Kensington: New South Wales University Press.

Wilson, P. G., M. M. O'Brien, P. A. Gadek, and C. J. Quinn. 2001. Myrtaceae revisited: A reassessment of infrafamilial groups. Amer. J. Bot. 88: 2013–2025.

Wrigley, J. W., and M. Fagg. 1996. Australian native plants: propagation, cultivation and use in landscaping, 4th ed. Chatswood, N. S. W.: Reed Books.

APPENDIX

List of Australian HISPID Designations for Species of Austromyrtus, Gossia, and Lenwebbia

The HISPID (Herbarium Information Systems and Protocol for the Interchange of Data) designations are cited as they appear in Henderson (1997) and Hyland et al. (1999), followed by the scientific name assigned to the particular voucher.

Austromyrtus sp. (Annan River BH 25539RFK) = *Gossia lucida*.
Austromyrtus sp. (Bamaga B. P. Hyland 10235) = *Gossia bamagensis*.
Austromyrtus sp. (Blackall Range P. R. Sharpe 5387) = *Lenwebbia lasioclada*.
Austromyrtus sp. (Brookfield L. W. Jessup 155) = *Gossia punctata*.
Austromyrtus sp. (Byerstown Range G. P. Guymer 2037) = *Gossia sankowskiorum*.
Austromyrtus sp. (Cape Flattery L. J. Webb+ 13537) = *Gossia retusa*.
Austromyrtus sp. (Claudie River G. P. Guymer 2052) = *Gossia floribunda*.
Austromyrtus sp. (Cooloolabin G. P. Guymer 2128B) = *Austromyrtus glabra*.
Austromyrtus sp. (Danbulla L. S. Smith 10123) = *Gossia bidwillii*.
Austromyrtus sp. (Forty Mile Scrub G. C. Stocker 1785) = *Gossia bidwillii*.
Austromyrtus sp. (Gillies B. Gray 1484) = *Gossia grayi*.
Austromyrtus sp. (Goldsborough G. Sankowsky 1076) = aff. *Eugenia reinwardtiana*.
Austromyrtus sp. (Isabella Falls G. Sankowsky+ 959) = *Gossia byrnesii*.
Austromyrtus sp. (Leo Creek B. Gray 5244) = *Gossia grayi*.
Austromyrtus sp. (Lizard Island G. N. Batianoff 12183) = sterile material; identity unknown (perhaps a species of *Gossia*).
Austromyrtus sp. (Main Range P. R. Sharpe+ 4877) = *Lenwebbia lasioclada*.
Austromyrtus sp. (McIlwraith Range B. P. Hyland 11148) = *Gossia macilwraithensis*.
Austromyrtus sp. (Mt. Lewis G. Bray 831) = *Gossia lewisensis*.
Austromyrtus sp. (Spencer Creek P. I. Forster+ PIF13701) = *Gossia bidwillii*.

Austromyrtus sp. (Upper Mudgeeraba Creek N. B. Byrnes+ 4069) = *Lenwebbia prominens*.
Austromyrtus sp. (Windsor Tableland B. Gray 412) = *Gossia shepherdii*.

NUMERICAL LIST OF SPECIES

1. A. dulcis
2. A. glabra
3. A. tenuifolia
4. L. lasioclada
5. L. prominens
6. G. inophloia
7. G. pubiflora
8. G. bidwillii
9. G. acmenoides
10. G. gonoclada
11. G. hillii
12. G. lewisensis
13. G. punctata
14. G. dallachiana
15. G. grayi
16. G. macilwraithensis
17. G. bamagensis
18. G. shepherdii
19. G. lucida
20. G. floribunda
21. G. myrsinocarpa
22. G. fragrantissima
23. G. sankowskiorum
24. G. retusa
25. G. byrnesii

INDEX TO NUMBERED COLLECTIONS

The numbers in parentheses refer to the corresponding species in the text and in the Numerical List of Species presented above. An plus sign (+) indicates additional collectors. In some cases the first initial of the collector could not be determined with certainty.

Altena, A. V. 3971 (4).
Bailey, F. M. 28 (9), BRI AQ-93327 (11), BRI AQ-93438 (11), BRI AQ-93441 (13).
Bailey, F. M. + MEL-726897 (6), MEL-726898 (6).
Bancroft, T. L. 99 (19).
Batianoff, G. N. 3196 (8), 3608 (8), 900337 (8), 940348 (8), 9406131 (8).
Batianoff, G. N. + 685 (8), 700 (8), 1368 (1), 3672 (8), 3675 (8), 3687 (8), 3772 (8), 3777 (8), 6119 (8), 11664 (8), 900144 (8), 900981A (7), 930513 (8), 940531 (8), 940539 (7), 940559 (7), 940569 (7), MS9108048 (8), 9403407 (7), 9404172 (8), 9405115 (7), 9405125 (7), 9609229 (8), 9812145 (8).
Bauerlen, W. 517 (8), 582 (8), 584 (8), 1363 (13), 1411 (8).
Baxter, P. 774 (1), 776 (1).
Bean, A. R. 1375 (8), 2021 (8), 2681 (2), 3899 (21), 5189 (8), 5308 (8), 5680 (21), 5933 (11), 5935 (6), 6097 (4), 6216 (11), 6501 (11), 6683 (9), 7077 (2), 7113 (4), 7133 (11), 7141 (8), 7147 (4), 7971 (22), 8010 (1), 8867 (11), 10178 (6).
Bean, A. R. + 5680 (21).
Beesley, P. + 824 (11), 875 (10), 981 (8), 989 (8), CBG8604325 (8).
Betche, E. NSW-144653 (3), NSW-144675 (1).
Bird, L. BRI AQ-437283 (11), BRI AQ-331061 (11), BRI-431238 (13).
Bird, L. + BRI AQ-431242 (8), BRI AQ-431244 (11).
Blake, S. T. 3325 (8), 7092 (1), 15469 (8), 18522 (8), 19123 (5), 22696 (1).
Blakely, W. F. NSW-144670 (3).
Boorman, J. NSW-144645 (3).
Bostock, P. D. + 820 (4).
Brown, R. 1802-5 (3), MEL-238768 (3), NSW-180701 (3).
Brushe, J. + 17 (8).
Burgess, C. 11046 (3).
Byrnes, N. 4057 (22), 4067 (22).
Byrnes, N. + 3949 (22), 4057 (22), 4062 (6), 4069 (5).
Cameron, E. 2220 (24).
Camfield, J. H. NSW-144658 (3).
Carr, T. 52 (23).
Carter, W. R. 1 (8).
Champion, I. G. + 1005 (7).
Cheel, E. NSW-144659 (3).
Cheel, E. + NSW-144648 (3).
Clarkson, J. R. 4219 (23).
Clarkson, J. R. + 6775 (23), 6901 (8), 8510 (20).
Clemens, M. S. BRI AQ-93430 (11), BRI AQ-93431 (11), BRI AQ-93433 (11), BRI AQ-93452 (6), MEL-517085 (11).
Constable, E. F. 22349 (5), 22358 (5).
Cooper, W. + 1120 (14).
Coveny, R. 4638 (1), 10897 (8).
Coveny, R. + 11438 (3), 12101 (3).
Cowley, E. 34 (21).
Crawford, A. R. (MEL 60615) (9).
Cribb, A. B. BRI AQ-331198 (7).
Crome, F. 231(8), 306 (8), 470 (21), 483 (14), 914 (18).
Cumming, R. J. 9342 (8), 10480 (8).
Cunningham, A. 49 (3).
Curdie, MEL-60227 (24).

Dallachy, J. MEL-67165 (14), MEL-67166 (14), MEL-92359 (21), MEL-92357 (8), MEL-92359 (21), MEL-92374 (21).
Dalliston, C. CC148 (20), CC243 (20), CC464 (20).
Dansie 20172 (12).
Dansie, S. J. 2013 (14).
Deane, H. NSW-144650 (3).
Degener, O. + 35,878 (14).
Dent, S. BRI AQ-584391 (3).
Dietrich, A. BRI AQ-93444 (10), 209 (10), 1307 (8), 1354 (8), 1934 (8).
Dockrill, A.W. 12 (8), 512 (8), 762 (20), 875 (20).
Dovey, L. G. 81 (8), 157 (8).
Dunn, C. J. + 112 (1).
Dunn, W. 231 (9), 4995 (9).
Durrington, L. 206 (1), 676 (1), 1197 (1).
Eaves, MEL 72380 (210).
Epps, F. C. 55295 (8).
Everist, S. L. 7733 (1), 7752 (1).
Fawcett, MEL-726906 (1).
Fell, D. G. (Note: All collections of D. G. Fell are prefixed with the initials DG or DGF.) 750 (8), 1415 (16), 1637 (23), 3044 (17), 4317 (19).
Fell, D. G. + 1212 (18), 2282 (14), 2427 (16), 2848A (19), 2863 (16), 2937 (8), 3019 (23), 3112 (17), 3155 (24), 3219 (23), 3231 (19), 3256 (23), 3277 (23), 3433 (17), 3438 (20), 3485 (24), 3518 (23), 3832 (17), 3833 (16), 3891 (20), 3954 (20), 3999 (20), 4041 (20), 4095 (20), 4124 (17), 4201 (20), 4255 (20), 4296 (20), 4361 (20), 4400 (20), 4498 (19).
Fensham, R. J. 261 (8).
Fitzalan, E. F. A. MEL-92366 (8), MEL-92367 (8), MEL-92370 (8), MEL-238843 (8), MEL-726880 (8), MEL-726883 (8), MEL-726884 (8), MEL-726915 (8), NSW-144689 (11).
Fitzsimon, D. 166 (18).
Flecker, H. 6329 (8), 3313 (21), 6954 (21), 8851 (21), 10795 (21).
Floyd, A. G. (Note: Floyd's collections are normally prefixed with AGF.) 819 (13), 1030 (22), 1031 (8), 1471 (1), 1630 (22), 1672 (1), 1936 (5), 1941 (9), 2007 (8), 2012 (9), 2146 (9).
Ford, A. 648 (3), 1827 (23), 1868 (23), 2021 (16), 2029 (11), 2124 (24), 2542 (23).
Ford, A. + 2562 (8), 2645 (19), 2671 (16), 2672 (16), 2674 (20), 2703 (8), 3073 (18), 3097 (11), 3176 (18), 3227 (15), 3288 (21)
Forster, P. I. (Note: all collections of Forster are prefixed with the initials PIF.) 2859 (8), 3323 (8), 3388 (8), 3769 (8), 6317 (20), 7226 (8), 7317 (8), 9640 (8), 13771 (14), 13774 (8), 13787 (8), 13882 (8), 13972 (8), 13989 (8), 14151 (8), 14294 (8), 14348 (4), 14392 (20), 14443 (21), 14447 (8), 14467 (8), 14625 (8), 14636 (9), 14662 (19), 14682 (8), 14723 (4), 14728 (16), 14729 (20), 14814 (4), 15057 (2), 15085 (4), 15179 (2), 15190 (11), 15237 (1), 15242 (8), 15524 (21), 16124 (11), 17059 (24), 19364 (24), 20144 (8), 20203 (8).
Forster, P. I. + 3248 (8), 4729 (8), 4758 (8), 4879 (8), 4992 (8), 5042 (13), 5178 (7), 5223 (8), 5413 (19), 5420 (19), 5563 (23), 8044 (8), 8292 (21), 10209 (17), 10228 (8), 10357 (20), 10403 (20), 10662 (4), 12475 (9), 12515 (8), 12530 (8), 12573 (9), 12609 (8), 13307 (9), 13411 (8), 13432 (23), 13461 (20), 13483 (20), 13498 (20), 13564 (20), 13567 (14), 13586 (17), 13641 (20), 13671 (18), 13701 (8), 13915 (1), 14072 (8), 14519 (9), 14523 (8), 14563 (9), 14609 (11), 14610 (11), 14614 (5), 15113 (10), 15114 (5), 15124 (4), 15292 (17), 15299 (20), 15489 (21), 15535 (18), 15536 (12), 15543 (12), 15547 (4), 15604 (19), 15612 (17), 15613 (12), 15615 (12), 15674 (5), 15692 (13), 15694 (4), 15832 (9), 15870 (6), 16119 (22), 16394 (5), 16647 (8), 17161 (21), 17229 (8), 17654 (4), 17685 (13), 18039 (4), 18362 (21), 19250 (23), 21877 (4).
Fosberg, F. R. 61619 (8).
Francis, E. 154 (8).
Francis, W. D. BRI AQ-93569 (18).
Franks, A. J. + AJF9209009 (1).
Fraser, L. NSW144671 (3).
Fraser W. 138 (11).
Geckeler, A. + 138 (8).
Gibson, N. 159 (8), TOI897 (8).
Gilbert, B. 25 (8).
Gittins, C. H. 14 (3).
Godwin, M. C3256 (8), C4108 (16).
Godwin, M. D. C3256 (8).
Goldsack, R. 719 (3); 720 (3); 721 (3).
Gordon, D. M. 118 (1).
Gray, B. 233 (21), 352 (14), 362 (4), 380 (15), 380 (15), 407 (18), 412 (18), 604 (15), 632 (18), 831 (12), 895 (21), 926 (4), 1205 (18), 1373 (4), 1462 (15), 1484 (15), 1484 (15), 1503 (12), 1509 (15), 1510 (15), 1518 (15), 1558 (12), 1596 (23), 1674 (21), 2237 (20), 2319 (8), 2481 (4), 2781 (8), 2844 (12), 3100 (21), 3272 (21), 3809 (21), 4275 (20), 4281 (20), 4706 (4), 5244 (15), 5999 (21), 6337 (23), 6530 (23), 6580 (19), 6847 (24), 7255 (14).
Grimshaw, P. G161 (22), G162 (9), G173N (13), G176N (9), G249 (11), G255 (13), PG809 (6), BRI AQ-546691 (10).
Grimshaw, P. + G201 (9), G206 (11), G320 (8), G342 (10), PG809 (6), PG2206 (11).
Gulliver, I. MEL-726885 (8).
Guymer, G. P. 297 (1), 1125 (8), 1136 (8), 1721 (7), 1991B (1), 2004 (7), 2023 (4), 2036 (23), 2037 (23), 2040 (21), 2047 (18), 2052 (20), 2054 (21), 2062 (14), 2065 (14), 2071 (8), 2072 (21), 2076 (8), 2078 (7), 2079B (8), 2079A (8), 2082 (9), 2089 (9), 2090 (13), 2103 (4), 2105 (5), 2111

(5), 2119 (6), 2121 (22), 2123 (8), 2126 (11), 2127 (8), 2128A (1), 2128B (2), 2129 (2), 2130 (6), 2133 (11).
Guymer, G. P. + 1791 (11), 1805 (9), 1806 (13).
Halford, D. Q776 (8), Q1829 (8).
Halford, D. + Q1829 (8), Q2170 (11), Q2877 (10), Q2879 (10).
Hamilton, A. A. NSW-144656 (3), NSW-144669 (3).
Harden, G. J. + 81298 (22).
Harrold, A. G. 290 (1).
Hauser, J. BRI AQ-424470 (10).
Hayes, H. C. 4 (13).
Hayes, H. C. + 2636 (5).
Hemsley, 7109 (8); 7117 (8).
Henderson, R. J. + H3043 (6).
Hill, G. F.(?) MEL-726445 (13).
Hind, P. D. 898 (3), 2203 (3), 5565 (3).
Hockings, F.D. 2 (8), BRI AQ-250142 (18).
Hoy, D. 141 (8).
Hunter, J. 336 (14); 920 (14); 1197 (21); 1531 (4); 3834 (18); 3885 (21); 4108 (18).
Hyland, B. (Note: Some collections of Hyland also bear the letters RFK, which are vouchers for the "Rain Forest Key" [Hyland + 1999]. Some herbaria use these letters as a prefix to the numerical designations, others as a suffix.) 535 (15), 1562 (18), 1615 (21), 1620 (15), 1645 (18), 1769 (21), 1941 (21), 2303 (21), 2334 (8), 2893 (12), 3256 (18), 3259 (16), 3322 (16), 3406 (18), 3418 (21), 3424 (14), 3970 (21), 3991 (4), 4070 (7), 4279 (8), 4341 (8), 4706 (14), 4988 (14), 5314 (4), 5365 (14), 5607 (14), 5873 (8), 6578 (4), 7415 (24), 7426 (20), 7481 (24), 7536 (15), 7584 (20), 7630 (16), 7774 (4), 8135 (23), 8989 (24), 9010 (24), 9345 (21), 9629 (20), 10089 (18), 10126 (16), 10211 (4), 10235 (17), 10289 (20), 10615 (24), 10869 (20), 11148 (16), 11637 (8), 11693 (12), 12096 (8), 13102 (20), 13952 (8), 13953 (8), 13954 (8), 14035 (19), 14095 (14), 14236 (14), 14686 (23), 15047 (8), 15048 (21), 15531 (23), 25285 (8), 25370 (8), 25385 (12), 25392 (18), 25982 (24), 25985 (24).
Irvine, A. 1809 (21).
Irvine, A. K. 537 (15), 596 (19), 761 (14).
Jackes, B. 9001 (8).
Jacobs, E. G. NSW-144647 (3).
Jago, 389 (21).
James, T. A. + 1219 (3).
Jensen, R. 799 (23).
Jessup, L.W. (Note: some collections of Jessup bear the prefix GJM or GJD.) 190 (11), 155 (13), 178 (13), 190 (11), 267 (8), 269 (8), 274 (8).
Jessup, L.W. + 20 (12), 208 (9), 218 (9), 572 (9), 754 (12), 875 (21), 1105 (21), 1198 (18), 1241 (18), 1269 (21), 1286 (15), 1338 (21), 1429 (21), 1975 (14), 2295 (18), 2354 (21), 2373 (14), 2483 (21), 2883 (19), 3024 (20), 3039 (20), 3040 (20), 3055 (19), 3115 (21), 4062 (6), 5168 (12), 5214 (18), 5215 (14), 5216 (21).
Johnson, L. A. S. 26 (3), 139 (5), NSW-144649 (3), NSW-144652 (3).
Johnson, L. A. S. + 47037 (5), NSW-144688 (5).
Johnson, S. 1401A (21); MEL-221899 (8), MEL-221900 (8), MEL-726889 (4), MEL-726890 (4), MEL-726893 (4), MEL-726894 (4), MEL-726896 (4).
Jones, D. L. 2265 (20).
Jones, W. T. BRI AQ-093425 (13).
Kajewski, S. F. 1053 (21), 1086 (14), 1343 (21), 1382 (4).
Kanis, A. 2133 (7).
Kemp, J. + 1903 (8).
Kenny, F. H. BRI AQ-93498 (21).
King, J. NSW-144666 (3).
Le Cussan, J. 299 (17).
Lebler, B.A. + BRI AQ-146957 (6), MEL-238757 (6).
Leichhardt, L. NSW-144714 (8).
Leiper, G. 5 (5), BRI AQ-414081 (10), BRI AQ-431237 (20), BRI AQ-431639 (11), BRI AQ-437282 (11), BRI AQ-437294 (20), BRI AQ-501219 (20), BRI AQ-509451 (20), BRI AQ-584408 (4), BRI AQ-637289 (20).
Leiper, G. + 2 (13).
Lyons, C. 55 (21).
MacDonald, W. 28 (8).
Maiden, J. H. 11897 (8), NSW-144655 (3), NSW-144667 (3).
Maiden, J. H. + NSW-144706 (22), NSW-144676 (1).
Martin, J. + 3002 (14).
McAuliffe, D. J. 918/15 (8), 7559/14 (8).
McBarron, E. J. 8611 (3), 14454 (3), 14898 (3), 17521 (3).
McCabe, J. BRI AQ-394774 (6).
McDonald, T. J. + 1408 (8).
McDonald, W. J. 3436 (9), 4152 (8), 4153 (8), 6174 (13), 6214 (9).
McDonald, W. J. + 2474 (5), 2813 (11), 4130 (8), 4136 (8), 4533 (7), 5621 (7), 5681 (8), 5792 (8), 5846 (8), 5882 (7), 5997 (21), 6031 (11).
McIlay, J. BRI AQ-584934 (10).
McKee, H. S. 566 (3).
McLachlan 12 (8).
McPherson, K. A. 82 (7), 87 (8).
Michael, N. 1476 (7), 1497 (7), 2256 (1).
Moore, C. 60 (8), 157 (1), MEL-60220 (24), MEL-726908 (1).
Moore, C. W. E. 2153 (3).
Moriarty, V. K. 1155 (8), 1979 (4), 2084 (15), 2622 (18), 2571 (12).
Morton, A. AM1007 (24), 1125 (24).
Nella, S. 2 (20), 5 (20), 6 (20).
O'Farrell, J. 84 (8), 100 (4), 110 (21).
Olsen, M. BRI-AQ 398234 (5).
Pearson, S. G. SP561 (21), BRI AQ-093434 (11).

Perry, F. 2 (21), BRI AQ-473230 (21), BRI AQ - 632036 (7).
Persieh, W. A. 58 (19).
Perseitz 298 (19), MEL-238760 (19), MEL-238761 (19).
Phillips, S. P. BRI AQ-569021 (6), 202 (13).
Playford, J. MEL-717893 (10), BRI AQ-569437 (10).
Pollock, A. B. + 126 (8).
Priest, E. D. 3 (11), BRI AQ-381026 (11), BRI AQ-457915 (11), BRI AQ-459431 (11).
Pullen, R. 7510 (20).
Risley, T. S. 32 (21), 65 (8), 367 (14), 509 (11).
Rodd, A. 3194 (8), 4201 (8), NSW-1446 (3).
Rodd, A. N. + 3519 (3).
Rothwell NSW-144704 (9).
Salasoo, H. 2686 (3).
Sandercoe, C. C1295 (2).
Sanderson, K. 223 (11), 224 (11), 257 (11), 261 (8), 320 (15), 460 (12), 481 (19), 779 (16), 1036 (12), 1086 (12), 1087 (15), 1488 (18), 1658 (11), 1792 (12).
Sankowsky, G. + 552 (14), 583 (14), 846 (4), 959 (25), 1102 (20), 1208 (17), 1241 (23), 1273 (19), 1295 (8), 1301 (20), 1304 (23), 1305 (23), 1306 (23), 1307 (23), 1309 (20), 1355 (4), 1364 (20), 1372 (20), 1377 (23), 1383 (23), 1408 (20), 1537 (21), 1554 (21), 1579 (19).
Sayer, W. A. 34 (4), 80 (4).
Schodde, R. 3300 (21), 3300 (21), 5581 (8).
Sharpe, P. R. 84 (21), 180 (2), 2028 (1), 3127 (1), 4144 (8), 4178 (7), 4186 (7), 4283 (1), 4287 (1), 4288 (1), 4312 (2), 4314 (2), 5387 (4).
Sharpe, P. R. + 4040 (7), 4439 (8), 4624 (8), 4836 (2), 4877 (4), 4918 (11), 4924 (2), 4988 (4), 5395 (4), 5399 (4).
Shirley, J. BRI AQ-93449 (6).
Simmonds, J. H. 186 (9), 187 (13), BRI AQ-93328 (13), NSW-144713 (8).
Simon W16 (9).
Smith, L. S. 3184 (14), 3211A (21), 3213 (15), 3215 (21), 3580 (8), 3691 (4), 3742 (8), 3748 (21), 3868 (8), 4139 (11), 4516 (8), 4516A (8), 4517 (8), 4659 (21), 4730 (17), 4741 (18), 4742 (14), 4756 (8), 4756A (8), 4787 (9), 5104 (1), 5352 (18), 5382 (9), 9848 (8), 10123 (8), 10137 (4), 10651 (19), 10729 (21), 11811 (17), 11870 (20), 13899 (5), 14350 (21), 14350 (21), 14414 (19), 14428 (21), 14531 (8), 14590 (7), 14635 (4), BRI AQ-375338 (11).
Smith, L. S. + 3620 (5), 3691 (4).
Snow, N. 7729 (4), 7733 (3), 7734 (5), 7737 (20), 7745 (9), 7746 (20), 7747 (11), 7748 (18), 7749 (22).
Snow, N. + 7452 (6), 7457 (1), 7459 (13), 7463 (5), 7466 (5), 7471 (13), 7740 (10), 7741 (9), 7742 (11), 7750 (4), 7752 (4), 7754 (6), 7755 (9), 7757 (9), 7759 (9), 7769 (9).
Sparshott, K. M. + KMS888 (11).
Stanton, J. P. 1022 (21).
Starkey, N. J. 49 (20), 50 (20).
Stephens, G. GBS9 (10).
Stephens, K. + ESK72 (8).
Stocker, G. 996 (18).
Stocker, G. C. 905 (14), 1460 (1), 1618 (12), 1785 (8), 1785 (8).
Strach, E. NSW-144651 (3).
Tardent, J. L. 231 (18).
Telford, I. R. 916 (8), 948 (8), 3031 (1), 3743 (1), 4353 (1), 7221 (3).
Telford, I. R. + 9011 (6), 9395 (19), 11139 (8).
Thomas, G. 46 (6).
Thomas, G. + RFR14 (10), RFR35 (6), RFR39 (10), RFR40 (22), 78 (11), 126 (11), RFR137 (11), 309 (10), 341 (4), BRI AQ-646359 (6), BRI AQ-636361 (10).
Thompson, E. J. MOR70 (11), 126 (11), MOR475A (4).
Thorne, R. F. + 21208 (2); 21408 (1), 27593 (8); 27641 (14).
Thorsborne, A. + 478 (8), 491 (8).
Todd, E. NSW-144712 (8).
Tracey, J. G. 14074 (20), 14167 (16), 14182 (23), 14229 (24), 14230 (14), 14365 (20), 14384 (20), 14394 (17), 14421 (17), 14434 (19), 14475 (20), 14499 (20), 14757 (16), 15348 (1), 15357 (11).
Tucker, M. 45 (12), 52 (8), 85 (19).
Tucker, M. C. (Note: collections typically prefixed with MCT.) 69 (21), 95 (14), 116 (21), 122 (4).
Unwin, G. L. 370 (15), 404 (16).
Vink, W. 17707 (8).
Volck, E. R185 (8), 2630 (17), QF57104 (15).
Walker, B. AGF1619 (11), AGF2006 (22), AGF 2007 (8), BRI AQ-345399 (11).
Webb, L. J. 999 (14), 2342 (8), 6373 (1), 6578 (14), 7121 (20), 7267 (21), 7345 (11), 7559 (20), 10057 (8), 14394 (17).
Webb, L. J. + 10A (24), 11B (24), 18A (17), 2163 (13), 5707 (21), 5730 (21), 6273 (18), 6373 (1), 6578 (14), 6722 (15), 6723 (18), 6776 (21), 6880 (21), 7165 (21), 7213 (15), 7253 (18), 7265 (17), 7272 (8), 7338 (17), 7634 (14), 7672 (17), 7673 (8), 7803 (14), 7961 (15), 8053 (11), 8054 (8), 8146 (8), 9143 (17), 9236 (14), 9261 (21), 9271 (16), 9444 (17), 9574 (16), 9575 (14), 9717A (24), 10057 (8), 10360 (8), 10390 (8), 10477 (21), 10526 (15), 10578 (21), 10589 (8), 11012 (24), 11042 (23), 11448 (14), 11709 (18), 11881 (19), 11953 (21), 12127 (21), 13537 (24), 13538 (24), 13733 (19), 13855 (20).
Wedd, J. 1106 (8).
Whaite, T. + 3495 (1), 3034 (5).
Whiffin, T. + 631 (14).
White, C. T. 16 (9), 25 (9), 1574 (14), 1856 (1), 6506 (1), 7319 (8), 9246 (8), 9592 (11), 9666 (1),

10340 (3), 10613 (21), 11071 (6), 12075 (6), 12455 (8), 12551 (9), BRI AQ-93437 (13), BRI AQ-93439 (11), BRI AQ-93447 (6), BRI AQ-319176 (18), BRI AQ-319177 (18), MEL-726888 (6).
White, K. J. 905 (18), 1180 (20).
Wiecek, B. + 408 (1).
Wilcox NSW-144699 (9), NSW-144700 (9).
Williams BRI AQ-345006 (22).
Williams, J. B. 2351 (5), NE43792A (11), BRI-517575 (11).
Williams, K. A. 75123 (8), 80129 (5), 83005 (8), 83010 (8), 83028 (8), 83050 (11), 87000 (5).
Wilson, P. G. 129 (3), 1347 (5), 13677 (3).
Woolls, W. MEL-238766 (3).
Worthington, M. 1456 (8).
Wrigley, J. + NQ1159 (21), 22288 (3).
Wyatt, R. A. 8 (8),
Young, P. 1748 (4).
Young, P. + 229 (11).

INDEX TO SCIENTIFIC NAMES

Accepted names are in roman type; the main entry for each is in **boldface**. Synonyms are in *italics*.